West Country Cruising

A yachtsman's pilot and guide to ports and harbours
from the Exe to Land's End and North Cornwall
including the Isles of Scilly

Mark Fishwick

Cover: Main picture: Fowey harbour, looking towards Polruan from the Hall Walk.
Lower inset picture: New Grimsby, Tresco, Isles of Scilly.
Back cover: Visitors' yacht haven, Falmouth

Edited by Geoff Pack
Design by Simon Firullo

Charts drawn by Mike Collins

Aerial photographs by Patrick Roach and South Devon Aerial Picture Library
All other photographs by the author

Published by
YACHTING MONTHLY
IPC Magazines Ltd
King's Reach Tower
Stamford Street
London SE1 9LS
1995

First Edition 1988, revised 1989
Third Edition 1995
© Copyright IPC Magazines Ltd and Mark Fishwick, 1995

ISBN 1-85277-101-1

Caution
Whilst every effort has been made to ensure the accuracy of the contents of this book, readers are
nevertheless warned that any of the information contained may have changed between the date when the
author carried out his research in 1994/1995 and the actual date of publication

Neither the publisher nor the author can accept responsibility for errors, omissions or alterations in this book.
They will be grateful for any information from readers to assist in the update and accuracy of the publication

Readers are advised at all times to refer to official charts, publications and notices. The charts contained
in this book are sketch plans and are not to be used for navigation. Some details have been omitted for the sake of clarity
and the scales have been chosen to allow best coverage in relation to the page size

Printed in Hong Kong through World Print

West Country Cruising

Preface to the third edition

In the five years that have elapsed depressingly fast since the last edition of West Country Cruising there have inevitably been many changes, most of a minor nature, although some of greater navigational importance, such as the new lights at the entrance to Salcombe.

Major new developments have been thin on the ground, with just one new marina at Clovelly Bay in Plymouth. However, many of the existing marinas have continued to consolidate their development, notably the major extension of Mayflower Marina in Plymouth and the completion of many other long awaited shoreside amenity buildings such as those at Falmouth, Brixham, Dart Marina and Darthaven Marina. Harbour Authorities, too, have continued to upgrade their facilities for visitors with new pontoons, moorings, much needed shower blocks and fuel berths.

However, aficionados of West Country Cruising will be quick to spot that the biggest change is within the book itself – in response to many requests from readers this now includes a major new section on cruising the Isles of Scilly, which will, I hope, serve as a useful introduction to this enticing but often demanding area.

As always I have done my best to ensure accuracy but apologise in advance for any mistakes or omissions that have crept in. Changes are continual and any assistance from readers to keep this book up to date will always be much appreciated. To those who have generously helped in the past, my thanks again.

Mark Fishwick

3 Gyllyng Street
Falmouth
Cornwall TR11 3EJ
February 1995

CONTENTS

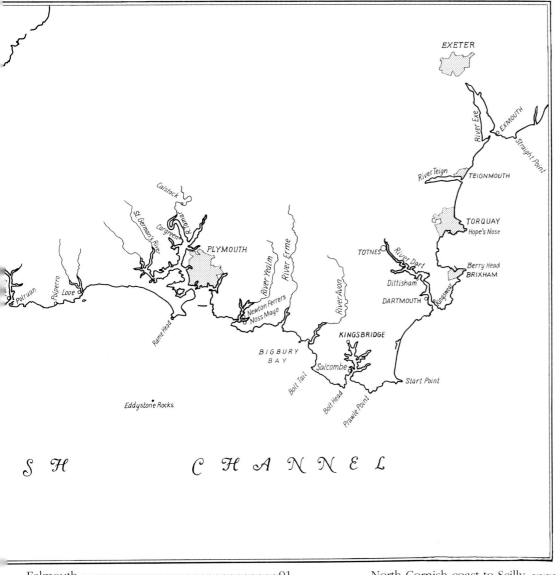

(map labels)

EXETER

River Exe

EXMOUTH

Straight Point

River Teign TEIGNMOUTH

Calstock

St German's River

Cargreen

R. Tamar

PLYMOUTH

River Yealm

River Erme

River Avon

TOTNES

River Dart

TORQUAY

Hope's Nose

Berry Head

BRIXHAM

Dittisham

DARTMOUTH

Kingswear

Newton Ferrers

Noss Mayo

KINGSBRIDGE

BIGBURY

BAY

Salcombe

Bolt Tail

Bolt Head

Prawle Point

Start Point

Polruan

Polperro

Looe

Rame Head

Eddystone Rocks

S H   C H A N N E L

INTRODUCTION

If I was given a brief to design a perfect cruising ground the West Country would not fall too far short of it, and I would be little inclined to make many changes. Year after year the magnificent coastline of Devon and Cornwall has attracted more and more visiting yachts with its ports and rivers deep and mostly of easy access, and its rock bound coast steep-to, with few off-lying dangers.

If there could be one major change that we would all agree on it would inevitably have to be the weather, but there again, what would English cruising be without its unpredictability? Generally speaking, well in the track of the prevailing south-westerly air stream and its accompanying depressions, I have noticed in recent years that the early summer can often provide some of the better spells of weather, and also we can be lucky in the early autumn. There does, however, seem to be a tendency for far more unsettled weather, often a blow or two, in August; not in itself a great problem once the long haul across Lyme Bay is astern, for a whole variety of harbours and safe anchorages abound, all within an easy day sail of each other, and many of the deep wooded rivers threading far inland such as the Dart and the Fal provide days of interesting exploration in themselves.

Way back in 1957 *Yachting Monthly* produced a slim little book called 'West Country Rivers', written by D J Pooley, and it proved to be an invaluable guide and a constant companion in my first early explorations of my home waters back in the 1960s. Long since out of print, it was a

Between Bolt Head and Bolt Tail

very pleasant surprise when *Yachting Monthly* invited me to revamp the book in 1987, and, including port and passage notes, to expand it into this cruising guide. Inevitably, my researches revealed that in twenty years much had changed 'down west'; sailing, once the pursuit of the select few had become a big leisure business, and the changes reflected it. Once forgotten backwaters and creeks, the haunt of rotting hulks, whispering mud and seaweedy smells, had been cleared and dredged, and marinas had blossomed forth; anchorages that were easy of access under sail were diminishing fast and as moorings encroached into every available space a reliable auxiliary was no longer a luxury but a real necessity.

However, in spite of it all, the essential beauty and atmosphere of the West Country continues to survive the onslaught well, and the sailing is as good as ever. Ashore, the varied and colourful villages and towns provide all the facilities a cruising yacht might need from essentials like good ale and launderettes to shipwrights and riggers. There is much of historical interest, and with vast areas fortunately preserved by the National Trust, many lovely walks. There can be few visitors to these waters who sail away disappointed, and hopefully, this guide will help to make your cruise just that bit more enjoyable . . .

Mark Fishwick

The author

Born and brought up on the River Exe, Mark Fishwick began sailing at an early age and soon ventured further to explore the rivers, creeks and harbours of the West Country in his parents 18ft 6in Alacrity sloop *Vallette*. He moved to Cornwall in 1973 soon after buying his present boat, the 34 foot 1910 gaff yawl *Temptress*, and since then has enjoyed a varied nautical life involving fishing, charter

skippering in the West Indies and yacht delivery. Now based in Falmouth, he divides his time between writing and sailing, and is convinced that there are few places to match the West Country as a cruising ground.

Acknowledgements

Inevitably a book of this sort could not have been completed without a large amount of assistance and help from others, but none of it would have been achieved without Vicki who has again selflessly devoted many hours to the never ending search for launderettes, cashpoints and shops open on Sundays. Organiser supreme, she has spent even more time keeping track of the chaos that I invariably managed to create in my wake, and in spite of it still managed to undertake the chart corrections and proof reading.

It would take several pages to mention everyone by name, but working east to west, my thanks to the Harbour Masters and their staff, Captain Jack Nott, Reg Mathews, Captain Mowat, Captain Mike Wier, Captain Colin Moore, Captain Peter Hodges, Mike Simpson, Commander Anthony Dyer, E T H Webb, Bill Cowan, Captain Mike Sutherland, Captain Hugh Bowles, Captain David Banks, Captain Andy Brigden, Jim Stephens, Dennis Swire, Captain Martin Tregoning, Andrew Munson, Eric Ward, L Wilkinson, Captain John Hinchliffe, Keith Graham, Laurie Terry and Henry Birch, Tresco, and Terry Tyler, Bill and Steve, St Mary's.

Additional thanks to my parents, Cliff and Pat Fishwick, Chris and Geoff Legg, Ian and Rosemary Cowan, Danny Murphy, Neil Clark, M P Avis, Jack Pender, Andrew Bray, Patrick Roach, Chris Hart, Robin Page, John Holman, Derek Rowe, Frank Collins, Liz O'Hara, Nicola Mansfield, Chris Price, Andy Barber, Ed Williams-Hawkes, Mike and Pat Miller, John and Penny Crane, Graham and Joy Swanson, Alan, Dave, Karl and Adam of Marine Instruments, Falmouth and all the Club Secretaries, boatyards, marinas and other people who have responded to my requests for information.

Finally my special gratitude goes to Geoff Pack for ensuring the ongoing continuity of *West Country Cruising*, Mike Collins for the charts, Peter Nielsen for the sub-editing, Paul Gelder for his magnanimity and Simon Firullo, not only for the design, but for cheerfully enduring my dubious company for the many long hours spent together during the preparation of this book.

GENERAL INFORMATION

Emergencies: All rescue services in the area covered by this book come under the coordination of the Coastguard service at either **Brixham (Tel: 01803 882704/5)** or **Falmouth (Tel: 01326 317575)**, both of which maintain a 24 hour listening watch on **VHF Channel 16 and normally work on channel 67, 10 and 73.**

Buoyage: All buoyage within the area of this Pilot falls for the most part into IALA System A. **LATERAL MARKS**, defining extremities of channels, are red cans, with square red topmarks and red light to be left to port; green or black conical buoys with triangular topmarks and green lights, to be left to starboard, when proceeding in the direction of the main **FLOOD** stream. **CARDINAL MARKS** are used in conjunction with the compass, placed N, S, E, or W of a hazard, usually pillar buoys coloured with a combination of black and yellow, (see abbreviations) with quick flashing white lights, the significant feature being the double cone topmark. Two black cones pointing up means the best water lies to the North, two cones pointing down, pass to the South, two points up and down (diamond) pass to the east, and two points together (wineglass) pass to the west. Other marks likely to be encountered include **ISOLATED DANGERS**, black double spheres topmark, as on Black Rock, Falmouth. **SPECIAL MARKS**, yellow buoys, often with a yellow X topmark, mark the limits of danger areas such as the firing ranges at the entrance to the River Exe. Note, however, in the upper reaches of some of the rivers that some buoys are privately maintained and do not necessarily conform to these shapes or colourings. A variety of beacons from simple poles or stakes in the mud, to complicated perches with wire stays and topmarks can mark narrower creeks.

Abbreviations: Colours of buoys, lights and their frequency is as follows:

R	Red
G	Green
Y	Yellow
B	Black
W	White
RW	Red and white
YB	Yellow and black (South cardinal)

Plymouth eastern breakwater end

BYB	Black, yellow, black (East cardinal)
BY	Black and yellow (North cardinal)
YBY	Yellow, black, yellow (West cardinal)
BR	Black and red
FR	Fixed red light
FG	Fixed green light.

Fl Flashing light, period of darkness longer than light. A number, indicates a group of flashes, eg. Fl(3). Colour white unless followed by a colour, eg Fl(3) R. Timing of whole sequence, including light and darkness, shown by number of seconds (sec or s) eg. Fl (3) R 15s. The range of the more powerful lights is given in Nautical miles (M) eg Fl (3) R 15s 25M

L.Fl Long flash, of not less than two seconds.

Occ Occulting light, period of light longer than darkness.

Iso Isophase light, equal periods of light and darkness.

Q Quick flashing light, up to 50/60 flashes per minute.

VQ Very quick flashing, up to 120 flashes per minute.

Mo Light flashing a (dot/dash) morse single letter sequence, eg Mo (S)

Soundings: Are all in metres and tenths of metres reduced to Chart Datum, the level of lowest astronomical tide, LAT. HEIGHTS: are in metres, above mean high water springs, MHWS. Drying heights, underlined are above LAT. Distances are in Nautical Miles,(2000yds),cables,(200yds) or metres.

Chartlets: Are of necessity simplified to show the main basics a visitor will require and should always be used with caution. Green shading = dries LAT.Blue = up to 5m LAT. White = over 5m LAT. *A back up of current/corrected Admiralty Charts is essential.*

Bearings: Are True from seaward. Magnetic variation should be applied as shown on current Admiralty charts. Variation in most of the West Country lies between 06°25' W and 04° 50' W, 1994, decreasing about 8' to 10' annually.

Marine Radio Beacons: Marine and Aero Radiobeacons for the area covered by this book are as follows:

Name	Lat/Long	Ident	Freq kHz	Mode	Range
Round Island	49°58.7N 06°19.3W	RR (. - . - .)	298.5	A1A	150M
Lizard	49°57.6N 05°12.1W	LZ (. - . . - - . .)	284.5	A1A	70M
Portland Bill	50°30.8N 02°27.3W	PB (. - - . - . . .)	313	A1A	50M
St Mary's Scilly	49°54.8N 06°17.4W	STM (. . . - - -)	321.0	Non A2A	15M
(day only Mon -Sat)					
Penzance Heliport	50°07.7N 05°31.0W	PH (. - - . . . . .)	333	Non A2A	15M
(day only Mon -Sat)					
Plymouth	50°25.4N 04°06.7W	PY (. - - . - . - -)	396.5	Non A2A	20M
Berry Head	50°23.9N 03°29.6W	BHD (- . . . . - . .)	318.0	Non A2A	25M
Exeter	50°45.1N 03°17.6W	EX (. - . . -)	337.0	Non A2A	25M
(day only)					

Tides: The maximum rate at Springs is used throughout this book to give an indication of the worst (or best, if you're going with it!) rate you will encounter. Generally, along the coast of the West Country, streams run parallel to the shore, but set in and out of the bays, and allowance should be made for this when crossing them. Within the bays, and inshore, streams are generally weak but increase considerably in the vicinity of headlands where overfalls are often found. Streams are also much stronger within the confines of the rivers, and here it should be remembered that these are fed by the large gathering basins of the inland moors, Dartmoor, Exmoor, and Bodmin, and after any period of rain, the amount of flooding fresh water can

noticeably increase the rate of streams at up to 2 knots, particularly in the upper reaches. Barometric pressure will also affect predicted tidal heights; explore upper reaches of rivers only on a rising tide, and do not push your luck too close to high water at Springs, or you risk being neaped for a considerable time!

Rise and fall of the tide is not at a constant rate, and the old 'rule of twelfths' is always a good guide:
Rise or fall during the 1st hour ¹⁄₁₂ of range
Rise or fall during the 2nd hour ²⁄₁₂ of range
Rise or fall during the 3rd hour ³⁄₁₂ of range
Rise or fall during the 4th hour ³⁄₁₂ of range
Rise or fall during the 5th hour ²⁄₁₂ of range
Rise or fall during the 6th hour ¹⁄₁₂ of range

Spring tides occur a day or so after the full and new moon, approximately every 15 days. High water is usually 50 minutes later each day; low water approximately six hours after high water.

Devil's Point, Plymouth. Mayflower Marina beyond

Weather and forecasts: As in most places in the British Isles the weather in the West Country can be very changeable, and can deteriorate fast. However, with comprehensive forecasting available, and ports within easy reach of one another, there is little excuse for getting caught out, and it should be remembered if you do, due to the exposed nature of much of the coastline, and ground swell, the seas will be considerably larger than elsewhere in the Channel, particularly with wind against tide.

If it is possible to generalise, the weather in the early summer May and June can often provide a fine spell, ideal for fitting out and usually giving rise to all sorts of hopeful predictions that this one is going to be the long hot summer! July, alas, has a tendency to dispel such promise, and can often be very mixed. Although summer gales are rare, August can be a poor month, and I have noticed a distinct tendency for there to be at least one bad blow during it, such as the Fastnet disaster in 1979, and the tail of Hurricane Charley in 1986. September, too, can be very unsettled although occasionally, the much vaunted Indian Summer can occur, if only briefly, as in 1994.

The dominant feature of the weather is the prevailing south-westerly air stream, across the warm waters of the Gulf Stream which touches the far west, and the depressions it creates usually pass to the north of the area with associated trailing warm and cold fronts creating an unsettled but fairly predictable sequence of events, drop in barometer, wind backing to the south or south-east, increasing, light rain becoming heavy for several hours, and wind freshening further before veering more to the west when rain becomes steadier with poor visibility in the warm sector. With the passage of the cold front, the rain will stop, skies clear, barometer begins to rise, and the wind will then veer north-westerly and often blow quite hard with blustery showers. As these die out, if another frontal system is in the offing, the wind will start to back, but there is usually a good chance of at least 6-12 hours of reasonable weather before it arrives.

The alternative, when the Azores High manages to push far enough north to keep the depressions at bay, and deflect them much further to the north, should theoretically result in the fine weather. Often it does, ashore, but the easterly airstream has a tendency to be fresh at sea, and creates very hazy conditions, blowing hard during the day, but dying down more at night. In fine weather, localised land and sea breezes are far more frequent, often blowing fresh out of the river approaches in the late afternoon, but generally the winds ease considerably by evening.

Rain is likely at any time! June, however, is often one of the driest months, and although the skies may be frequently overcast and grey, heavy and prolonged rainfall is much less than in some parts of the country.

Fog, fortunately, is not very frequent this far west, but in warm weather sea mist can occur without warning, often towards the end of a fine warm day. In fine, settled weather, usually when high pressure prevails, early morning mist in rivers will often be thick but clears within a couple of hours. However, poor visibility associated with the passage of fronts is far more frequent, misty rain, good old Cornish 'mizzle'. Thunderstorms can occur but not particularly frequently, tending to generate over the continent and drift over the channel further to the east.

Forecasts can be obtained from a variety of sources
BBC Radio 4 (198kHz) Shipping Forecasts at 0033, 0555, 1355, 1750 daily, particularly for sea areas Portland, Plymouth, Sole, Lundy, Fastnet.
BBC Local Radio Stations also issue Coastal waters forecasts, but their times are subject to annual change.
BBC Radio Devon broadcasts on 990kHz, (Exeter area),1458kHz, (Torbay area), 855kHz (Plymouth area).
BBC Radio Cornwall broadcasts on FM 95.2, 630kHz (north and east) FM103.9, 630kHz, (mid and west), and FM 96.0 (Isles of Scilly).
BBC Radio 4 has Inshore Waters Forecast at end of daily broadcasting.
By telephone: Marinecall, Area 8, Hartland to Lyme Regis (Tel: 0891 500458). Plymouth Weather Centre: 01752 251860.
In addition, the following BT Coastal Radio Stations issue VHF bulletins at 0733 and 1933 hours UT:
Start Point Radio, VHF Channel 26. Pendennis Radio, VHF Channel 62. Land's End Radio, VHF Channel 27. (Channel 64 beamed to Scilly)
Conveniently, *SUBFACTS* and *GUNFACTS* are also broadcast at these times, giving details of Submarine and Gunnery activities in the area for the next 24 hours.

The Coastguard issue forecasts on VHF Channel 67 every four hours starting Brixham Coastguard 0050 and Falmouth Coastguard 0140. When strong wind or gale warnings are in force these will be broadcast every two hours from the starting times. The Coastguard will normally provide a repeat of the latest forecast on request, but this facility should not be abused. They will also provide the latest forecast by telephone: Brixham Coastguard (Tel: 01803 882704/5). Falmouth Coastguard (Tel: 01326 317575).

Yacht Clubs: The traditionally stuffy image of many Yacht Clubs has fortunately changed much in recent years and most are only too happy to welcome genuine cruising yacht crews on a temporary basis to use their facilities. However, this normally only applies to members of other recognised clubs, although some, such as the Salcombe Yacht Club and Dartmouth Yacht Club extend a welcome to all. Whatever the situation, it is a privilege that visitors should not abuse, and on arrival at any new Club immediately introduce yourself to the Steward, Secretary or a member, ascertain if you are indeed welcome and ask for the visitors' book which should be signed not only by yourself but also your crew. For two or three nights, this will be sufficient, but should you wish to use the Club facilities for a longer period, check with the secretary as to the arrangements for temporary membership which is often available. Remember at all times, you are their guest and behave accordingly.

Customs: As long as you are only moving between countries within the European Community there is no longer any need to make declarations to the Customs on departure from or re-entry into the UK. *Customs officers do however retain the power to search any boat at any time.* However, if you are arriving from or departing for any port outside the EC, **AND THIS INCLUDES THE CHANNEL ISLANDS**, a declaration will have to be made, and Q flag flown on re-entry into UK 12 mile territorial waters and you must contact the Customs immediately on arrival. Throughout the West Country all such calls to Customs are co-ordinated through one number: **PLYMOUTH 01752 220661**. The nearest local office will then be informed and Officers despatched to effect clearance.

Sadly, drug-smuggling has been increasing along the British coast and the West Country is no exception. Any suspicious activity, such as gear being unloaded from boats in isolated bays, or small craft alongside each other at sea, should be reported immediately to the Customs, dial **0800 59 5000** and ask for **DRUG SMUGGLING ACTION LINE.**

As increasing numbers of pleasure boats are being used in this insidious trade, ultimately it can only bring the name of genuine cruising yachtsmen into suspicion and disrepute, so it is very much in our interests to do anything we can to help assist in stamping it out. It may be irksome at times to be asked, often frequently, who you are and where you have come from by Customs patrols; try not to forget that they are only doing their job.

Passages
STRAIGHT POINT TO START POINT

Favourable tidal streams

Portland Bill	Bound West: HW Dover
	Bound East: 5 hours after HW Dover
Start Point	Bound West: 1 hour before HW Dover
	Bound East: 5 hours after HW Dover

Straight Point to Start Point

Passage charts for this section are:

BA: 2675 English Channel. 442 Lizard Point to Straight Point.

2454 Start Point to Needles. 3315 Berry Head to Bill of Portland. 1613 Eddystone to Berry Head

Imray: C5 Portland Bill to Start Point

Stanfords: 12 Needles to Start Point

French: 4813 Du Start Point a Needles

Radio beacons

Beacon Sign	Range (M)	Frequency (kHz)	Call
Marine beacons			
Portland Bill	50	313.0	PB
Aero beacons			
Exeter airport (daytime)	25	337.0	EX
Berry Head	25	318 .0	BHD

Coastal Radio: Start Point Radio, working channels 26,65,60. Traffic lists usually broadcast at 3 minutes past the hour. Weather forecasts at 0733 and1933 UT

'Going West?' said the skipper of the pilot cutter. 'Yes,' we answered, and felt like adventurers. 'And you?' He shook his head. We'd have a head wind, he reminded us, all across the bay. We knew it, but we had a good ship, too.

The west wind still blew. When we were clear of the harbour we backed the jib and let the boat lie while we hoisted the dinghy on deck, and lashed it. That done, we let draw, set the foresail and mizzen, and stood away for the Shambles Light.

Our voyage had begun. We were bound west, to visit a new country beyond the Bill . . .'

From Portland Bill to Berry Head it is exactly 40 miles, and 47 to the Start. It never seems less, and usually feels much further. The opening lines were written by Aubrey de Selincourt in 1948, and nothing has changed - the odds are still just as much in favour of being on the wind most of the way, plodding into the prevailing south-westerlies. There is always a certain nervous relief at watching Portland drop safely astern, but for many that is soon replaced by other anxieties as the long haul across Lyme Bay, in normal visibility, is often the first opportunity to make a passage out of sight of land .

There is at first, the pleasant anticipation of the cruising ground ahead, but as the hours pass, and there is still nothing to be seen, those first niggling doubts begin to set in. Is the compass really accurate? Did I allow for tidal set, and leeway? Always, an anxious eye on the weather, the slight greyness to windward, and the hint of an increase in the wind. And underlying it all, that hollow awareness that there are now no safe harbours under your lee.

But then, at last, there it is, no longer a figment of wishful thinking but, unmistakeably, the land again, and looking, of course, not the least bit as you imagined. A featureless, thin, low line, and no sight of Berry Head and Start Point, the two bold headlands you expected to see for they will be indistinguishable until much closer – Berry Head distinctively square and flat topped, and the Start, an unmistakeable jagged cockscomb, with a conspicuous white lighthouse and two very tall radio masts close by. In practice it is invariably much easier to make a landfall on a strange coastline at night, ideally just before dawn, for then the lights will take away the doubts. Start Point (Fl 3 10s) has a 25-mile range, and Berry Head (Fl 2 15s) has a

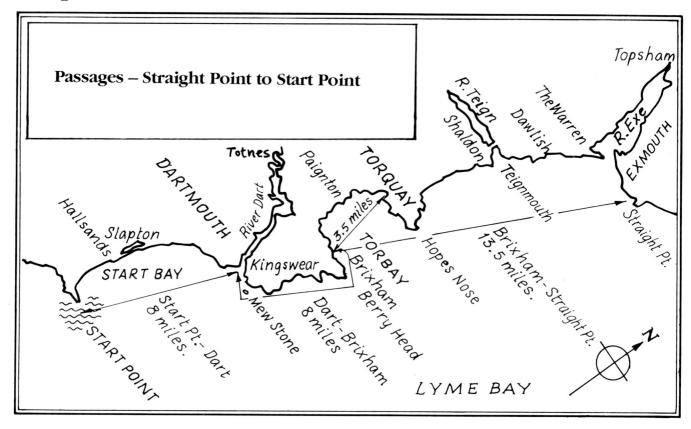

Passages – Straight Point to Start Point

14-mile range. For first timers across the bay, the prospect of a night passage is probably not very appealing and it is more likely that your approach will be a race against daylight, which is no real drawback, as your probable first ports-of-call, Torquay, Brixham or the River Dart are all well lit and easy to enter. Though tired, the excitement of landfall and arrival in the West Country should definitely carry you through.

The coastline you will now be cruising is varied, spectacular in places, and very different from that to the east of Portland. The last prominent chalk headland is at Beer, 15 miles east of the River Exe, and from here onwards the high cliffs of deep red Devon sandstone begin, broken only by the seaside towns of Sidmouth and Budleigh Salterton, where beaches of large round pebbles rise steeply from the sea. Inland, Woodbury Common, an area of heathland and forestry plantations, rises to nearly 200m in height.

The Exe estuary

Straight Point (Fl R 10s) forms the eastern side of the entrance to the River Exe, a low lying sandstone promontory, with two prominent flagstaffs indicating whether the rifle ranges are in use. Two yellow DZ buoys just under a mile south-east and east of the point mark the seaward limit of this range. Straight Point light is of seven miles range, visible from 246°- 071°T.

The **River Exe**, is a broad, drying estuary, navigable for six miles inland to Topsham, but less frequented by visitors, who are, perhaps, unnecessarily put off by the reputation of the bar. Although dangerous in strong east or south-easterly onshore winds, in favourable conditions the river is definitely worth a visit, and the entrance channel is well-buoyed, leading inwards from the 'East Exe' BYB east

cardinal buoy (Q (3) 10s) just under half a mile SW of Straight Point. Although well lit, the entrance is not recommended unless you have experienced it in daylight.

The high red cliffs of Orcombe Point lead in to the beach and town of **Exmouth** on the eastern side of the entrance. Opposite, the Warren, a long sandy promontory closes off the greater part of the wide mouth of the Exe and shallow banks extend nearly a mile to seaward. To the west, the red sandstone cliffs reappear between the resort towns of Dawlish and Teignmouth, and beneath them the main London/Penzance railway line enjoys a spectacular run along the coast. Inland, the land rises to over 250m again into the Haldon Hills. There are no off-lying dangers except Dawlish Rock, covered 2.1m LAT ½ mile ESE of the town, and a course ½ mile from the shore can safely be followed if you wish to admire the scenery, weathered cliffs, and sandstone pinnacles like the Parson and Clerk.

Inshore, the tidal streams along this stretch of coast are generally weak, and run parallel to the coast, the NNE going flood beginning 5 hours after HW Dover, and SSW ebb beginning just after HW Dover, attaining a maximum of 1 knot at Springs. Channelled in the closer approaches to the rivers, however, rates increase considerably, attaining in excess of 4 knots at Springs.

The Exe from south-east – DZ buoy and Straight Point

The Teign estuary

Immediately to the west of **Teignmouth**, which is easily located by the prominent church tower, and long pier on the seafront, there lies the narrow entrance to the River Teign, bounded on the western side by the prominent high sandstone headland of the Ness (QWRG) a sectored approach light with 7 mile range, not that any approach should be made at night by strangers. Here, too, a bar renders the entrance dangerous in onshore winds. Facilities for visitors within the river are severely restricted by a bridge which cuts off the upper reaches half a mile above the entrance. There is a yellow sewage outfall buoy (Fl Y 5s) 1.16 miles south-east of the Ness.

The coast remains steep from the Ness for the next four miles westwards, with no dangers except within a few boat's lengths of the shore. Topped with fields, trees, and isolated houses, the red sandstone gives way to pale grey and ochre limestone off Babbacombe Bay where there is a reasonable anchorage in westerly winds, and Anstey's Cove to the west of the quarry, scarred Long Quarry Point is another good temporary anchorage, although care must be taken to avoid three drying rocks near the southern entrance point to the cove.

Torbay

In contrast to the high cliffs to the east, Hopes Nose, the eastern boundary of Torbay, slopes gently down to the sea, grassy turf, with low rock ledges and cliffs at its base, the home of the largest kittiwake colony in Devon. Just under four miles away, the distinctive flat topped line of Berry Head marks the western limit of the bay, which takes a deep, sheltered bite into the Devon coast. Well protected in westerly winds it was a traditional anchorage for the Navy prior to the development of Plymouth, and it has always been a popular venue for yacht racing, graced by the magnificent 'Js' during their brief heyday in the 1930s, and more recently, powerboat racing, with the punishing Cowes/Torquay Race a long standing event.

To seaward, the low, flat Lead Stone and the 32m high Ore Stone, ½ mile offshore, are popular with the seabirds, too, and white with droppings. There is a deep passage between the two islands, but do not pass too close to the south-west of the Ore Stone where the 'Sunker' lurks awash at LAT. In calm weather, with an absence of swell, it is possible to land on the Ore Stone, anchoring just north of the island on the rocky ledge which has 3m LAT. Pick anywhere on the rocky shore, and lift your inflatable up onto the rock platform. I would not, however, recommend leaving your mother ship entirely unattended.

The last of the Hopes Nose islands is Thatcher Rock, south-west of the point, a jagged pyramid rising to 41m. Behind it, the concave sweep of the large hotel high on the cliffs is a foretaste of what will emerge as **Torquay** opens beyond the next point. Just under ½ mile west of Thatcher Rock lies Morris Rogue, a shoal with a least depth of 0.8m LAT. Keep the Ore Stone open of Thatcher Rock and you will clear this hazard, otherwise deep water extends safely right to Torquay Harbour entrance.

Berry Head from the south, Torbay opening in the distance. Note large ship waiting to pick up Pilot

Straight Point to Start Point

Although the actual harbour mouth is not easy to spot until close, Torquay is an unmistakeable proliferation of large buildings, tower blocks, with the large façade of the Imperial Hotel high on the cliffs just to the south of the harbour. Torquay is easy to enter, both day and night, there is a large marina with ample berthing for visitors and excellent facilities.

The whole of the coast backing Torbay is one large urban sprawl, the 'English Riviera' an unbroken line of busy beaches, promenades, resorts and holiday camps in the hinterland, comprising Preston sands, Paignton sands, Goodrington sands and Broad sands. Along their length yellow speed limit buoys are positioned about two cables offshore during the summer, with '5 KTS' on square top-marks, restricting the areas for swimming, and inside these controlled areas you should proceed with extreme caution.

Paignton, just under half way across the bay, is a popular resort with a traditional pier; west of this is the sandstone bluff of Roundham Head, and a small drying harbour is situated on its northern side. Given over completely to local moorings, and busy with tripper boats, the harbour is not recommended for visitors, except for very small boats which can sometimes find room to dry out against the walls. Rocky ledges extend eastwards from the south wall, marked by a red beacon, and an approach should be made from a north-easterly direction, close to HW. On arrival berth on the south-east wall and seek out the Harbourmaster to see if space is available. As there are heavy mooring chains across the harbour do not attempt to anchor. Facilities are limited, but there is a restaurant and pub beside the harbour and a small chandlery.

The once-peaceful anchorage of Elberry Cove in the extreme south-western corner of the bay is now marred by water skiing activity. East of it, the houses give way to the large expanse of Churston golf course, beneath it the cliffs, broken with many disused quarries, fall steeply into the sea with no dangers at their foot. Just west of the wide entrance to Brixham outer harbour is Fishcombe Cove, a quiet secluded anchorage in westerly winds. **Brixham**, a busy fishing port, is well sheltered by the long breakwater except in north-west winds when Torquay is the obvious alternative; facilities are good, and berths are available for visitors in the outer harbour.

Brixham to Dartmouth

With the exception of rocks extending a cable to the north of Shoalstone Point, the coast between Brixham and **Berry Head** is steep-to. The headland, 55m high, is an impressive sight, rising steeply with its northern side heavily quarried. A coastguard lookout with latticed radio mast stands beside the lighthouse at the eastern end which drops almost vertically into the sea, and though not recommended, you could sail to within a boat's length of the foot of the cliffs. Off the point the tide attains 1.5 knots at Springs, the north going flood begins 5½ hours after HW Dover, south going ebb 1 hour before, the streams running parallel to the coast.

Half a mile to the south of Berry Head, off Oxley Head which has a ruined fort and car park on its top, there are the Cod rocks, two steep islands, the larger of the two, East

Dartmouth from the east. Note Mewstone and daymark

Cod 9m high. Between them lie the drying Bastard Rocks, and although it is quite safe to enter the bay to the north to admire the impressive limestone cliffs overhanging a large cave, keep well to seaward when leaving, and do not attempt the inshore passage between the rocks.

Various dangers and shallows extend for about half a mile to seawards for the rest of the four mile passage along this stretch of coast, and a sensible offing should be maintained. Mostly high, rolling turfy hills, sloping to low irregular limestone cliffs, it is interspersed with several beaches and coves in steep valleys. Mag Rock, drying 0.3m, lies 1½ cables east of Sharkham Point, and to seaward, Mudstone Ledge, safe enough, with a least depth of 5.4m does, however, kick up an uncomfortable sea at times. It is also a popular spot for pots, and markers and buoys, the scourge of the whole Devon and Cornwall coastline, will be found in abundance from here on. Many are barely awash, and a careful lookout must be kept at all times, which renders inshore passages at night particularly wearisome if under power.

A course at least half a mile from the coast will clear both Druids Mare, a group of rocks drying 2.1m 1½ cables SSE of Crabrock Point, and Nimble Rock, a particularly insidious outcrop with 0.9m over it LAT, which lurks nearly two cables SSE of Downend Point. Start Point lighthouse open of Eastern Blackstone will clear the Nimble.

Eastern Blackstone, steep and 16.5m high, is not difficult to miss, nor too, the much larger, 35m high jagged outcrop of the Mewstone beyond it to the south-west. Do not be tempted to cut the corner from here into **Dartmouth** as submerged rocks extend nearly three cables to the south-west of the Mewstone. Inshore, high on Froward Point is the prominent Dartmouth day beacon, a tower with a wide base that is particularly valuable for locating the Dart when approaching from the south or south-west, as the narrow entrance to the river is very difficult to distinguish in the high folds of the coastline.

Start Bay

The wide sweep of Start Bay extends nearly 8 miles south-west from Dartmouth to Start Point, and inshore the high cliffs undergo a dramatic transformation into Slapton Sands, a long low beach enclosing the fresh water lake of Slapton Ley, a valuable wildlife sanctuary. During the War,

Start Point from the north-east. A distinctive and often demanding headland, it should always be treated with caution in poor weather

this area was used extensively by the American forces practising for the D-Day invasions. There are no inshore dangers, and with an offshore breeze a fine sail in calm water can be enjoyed along its length close to the shore.

West of Torcross, the cliffs begin to rise again, with the tiny fishing village of **Beesands** at their foot, and its less fortunate neighbour the ruined village of Hallsands. During the latter part of the nineteenth century nearly 700,000 tons of shingle were removed from this corner of the bay to build the new docks in Devonport — vital natural defences for the small fishing village, which met its fate on 26 January 1917 when an easterly gale and high tide destroyed the entire village, leaving just the one ghostly ruin that remains today. In settled offshore weather, if time permits, or waiting for the tide west around the Start, this is a handy anchorage, and a trip ashore has a distinctly ghoulish fascination.

Ironically, it is further offshore that Start Bay poses more problems, for it is here that you encounter the only real offshore bank in the West Country, the **Skerries**, which extends 3½ miles in a NE direction from a position 6 cables NE of Start Point. Although the depths over the bank are adequate for most small craft, the shallowest point being the SW end with a depth of 2.1m LAT, it creates dangerous breaking seas in bad weather, at which time it should be given a wide berth, and an approach to Dartmouth from the south not attempted until the red 'Skerries' can buoy is passed at the NE end of the bank.

Deriving its name from the Anglo-Saxon meaning, a

'tail', **Start Point**, is one of the more distinctive West Country headlands with five grassy hillocks topped with rocky outcrops about 60m high along its prickly spine, and two very high 264m BBC radio masts. Built in 1836, the white buildings and lighthouse (Fl 3 10s) perch neatly above the low cliffs at the eastern end, and a fixed red sector covers the Skerries Bank visible 210°-255°T.

Although weak in Start Bay, tidal streams now have to be reckoned with again, becoming much stronger off Start Point and attaining up to 4 knots at Springs, creating a race which extends about a mile to seaward from the lighthouse. Three miles south of the point the streams are weaker, just over 2 knots at Springs, the ENE flood beginning 5 hours after HW Dover, and the WSW ebb about 1½ hours before HW Dover. Closer inshore the tide turns about half an hour earlier. There is no inshore passage as such but in fair weather less turbulence can be found closer to the point, taking care to avoid Start Rocks, and the numerous pot markers. *In bad weather, especially wind against tide, the race should be taken seriously as it produces heavy overfalls, and the point should be passed* **at least** *two miles to the south.* As ships bound down Channel for Plymouth and Fowey close the land at this point there is a noticeable increase in traffic.

In fog or poor visibility, audible aids to navigation are restricted to the foghorn on Start Point (60s) and the bell on the Skerries Buoy.

RIVER EXE

Tides: HW Dover - 0445. Range: Exmouth MHWS 4.6m - MHWN 3.4m, MLWN 1.7m - MLWS 0.5m.
(HW Topsham approx 20 mins after HW Exmouth.) Spring ebb can attain 5 knots off Exmouth docks
Charts: BA: 2290. Stanford: 12. Imray: Y43, Y45
Waypoint: East Exe Buoy, 50°35' 97N 03°22'30W
Hazards: Pole Sands, Maer and Conger Rocks (lit by fairway buoys). River approach dangerous in strong east and southerly weather particularly on the ebb, near LW. Strong tidal streams in entrance. Large part of river dries

The less visited estuary of the Exe differs greatly from its deep, wooded companions further west, for it has a character far more in keeping with one of the rivers of the East Coast, and could easily be dubbed the 'Blackwater of the West'. Flat and wide, it forms an open expanse over six miles long at HW, which dries for the most part into banks of mud and sand, creating a habitat for the abundant bird life for which the estuary is internationally famous. As most yachts heading west aim to make a landfall on Berry Head or Start Point the Exe, lying inshore of the normal track across Lyme Bay, tends to be overlooked by visitors. However, the channel is well-buoyed and in the right weather it can provide a unique diversion from the better known cruising haunts. For many years this was my home patch. Though busier than it used to be, there are still several peaceful anchorages and interesting places to explore and it is unlikely that you will sail away disappointed with what you have found.

Approaches

The entrance to the river is a long, narrow channel running parallel to the beach and foreshore at Exmouth, flanked to seawards by the Pole Sands, an extensive area of shallow water that dries considerably. The tides can run hard, particularly on the ebb at Springs, and in strong winds from the south or south-east a first time visit to the Exe should not be considered, as a confusing area of breaking sea soon builds up in the approaches, and the run in along an uncomfortable lee shore leaves little room for mistakes. Ideally an approach should be made from the south-west, in favourable weather and on the flood, ideally 2 hours after local LW, when most of the drying banks can be seen. Although well lit, entry is not recommended at night without local knowledge. ***Due to the changing nature of the sands and buoyage within the Exe an up-to-date copy of Admiralty Chart 2290 is strongly recommended.***

Exmouth Bar is two miles to the east of the town, which is best located by the conspicuous tower of Holy Trinity Church, and, approaching from the south-west, the 'East Exe' BYB east cardinal buoy (Q (3) 10s) marking the entrance to the buoyed channel lies just under ½ mile SW of Straight Point. This low headland lies just south of the prominent white diamond shape of Sandy Bay caravan park with high red cliffs beneath it. There are two flagpoles on the headland, and in its vicinity there are invariably large numbers of small local angling boats. Straight Point is

used by the Royal Marines as a firing range and its seaward limits are marked by two spherical yellow DZ buoys to the south-east and east both (Fl Y 3secs). From March to October firing can take place on weekdays between 0800 and 1600, and the danger area should be avoided. When firing is in progress a safety boat patrols the area, and red flags are flown from the flagstaffs on the point. No flags mean no firing, a flag on each pole indicates firing in progress, and two flags on each pole means firing temporarily ceased. The range can also be contacted on VHF channel 08, call sign *Straight Point Range*.

From the East Exe buoy, the channel is well-marked with even numbered red cans to port and odd numbered green conical buoys to starboard, the first of which is red No 2. The shallowest water, the bar proper, lies between Nos 3 and 4 buoys south of the high red sandstone cliffs that form Orcombe Point, where there is a depth of 1.4m LAT. Exmouth beach and promenade run in close on your starboard hand, with brightly painted beach huts, and then a long row of typical Edwardian seaside hotels as you near the real mouth of the river, between the **Warren**, a long, low, sandy promontory, and the entrance to Exmouth Docks. Do not be tempted to stray from the channel, there is little room outside of it and the Pole Sands are steep-to and surprisingly hard if you run aground. Rocky ledges extend southwards between Nos 5 and 9 buoys, forming **Maer and Conger Rocks**, a notorious grounding spot, and it has been becoming increasingly shallow between Nos 4 and 7 buoys with barely 1m at Springs. Little more than

The East Exe buoy and distinctive sandstone cliffs

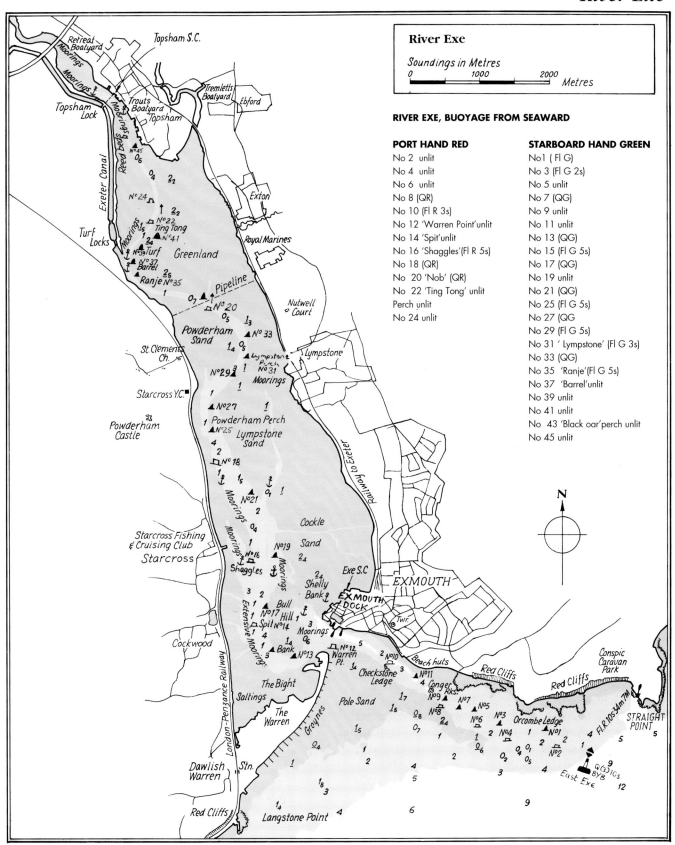

River Exe

Soundings in Metres

0   1000   2000

Metres

RIVER EXE, BUOYAGE FROM SEAWARD

PORT HAND RED	STARBOARD HAND GREEN
No 2 unlit	No1 (Fl G)
No 4 unlit	No 3 (Fl G 2s)
No 6 unlit	No 5 unlit
No 8 (QR)	No 7 (QG)
No 10 (Fl R 3s)	No 9 unlit
No 12 'Warren Point' unlit	No 11 unlit
No 14 'Spit' unlit	No 13 (QG)
No 16 'Shaggles' (Fl R 5s)	No 15 (Fl G 5s)
No 18 (QR)	No 17 (QG)
No 20 'Nob' (QR)	No 19 unlit
No 22 'Ting Tong' unlit	No 21 (QG)
Perch unlit	No 25 (Fl G 5s)
No 24 unlit	No 27 (QG)
	No 29 (Fl G 5s)
	No 31 ' Lympstone' (Fl G 3s)
	No 33 (QG)
	No 35 'Ranje'(Fl G 5s)
	No 37 'Barrel' unlit
	No 39 unlit
	No 41 unlit
	No 43 'Black oar' perch unlit
	No 45 unlit

100m wide, the channel is at its narrowest at this point. **Checkstone Ledge**, a rocky outcrop to the west of No 10 buoy is also one to avoid, and at LW do not be tempted to cut the corner towards the low sandy promontory of Warren Point as this is continually extending south-east, forming a large area of gravelly shallows. Leave No 12 buoy well on your port hand and hold a course towards the dock end before turning west across the river.

Exmouth

The tidal streams run strongly, in excess of 3 knots on the ebb at Springs, creating a confused tidal lop in the narrows, and a very strong run across the entrance to Exmouth Dock which is hidden until you draw abeam of it. Beware of the frequent passenger ferries and other boats emerging. Entered through a swing bridge, Exmouth Dock is privately owned and was regularly used by coasters until

17

The Exe entrance channel is well buoyed and runs in parallel to Exmouth beach and promenade, right. The sandy spit of Dawlish Warren can be seen on the distant left, with the Pole Sands just covering behind the departing yacht

1989, since when it has been undergoing redevelopment for use by pleasure craft. There are pontoons and alongside berths within the dock, with an average maximum depth of about 1.6m MLWS, but less in places. Although the berths are private, space can usually be found for visitors if berth owners are away, call *Exmouth Dock* on VHF channel 14, or berth alongside the pontoon on the starboard side of the dock entrance and enquire at the Dockmaster's Office on the quay. The overnight charge for a 9m boat works out at £6 per day plus VAT. If there is not space within the dock, they might be able to provide a mooring in the river. Although there are no facilities other than public lavatories, it is anticipated that toilets and showers will eventually be provided as part of the major building development planned for the surrounding area and due to begin in 1995.

The pool to the north-west of the dock entrance off 'The Point' is completely taken up with local moorings, and the Exmouth Lifeboat is also moored here. Anchoring is not

There are visitors' berths available in Exmouth Dock which is being developed as a marina

recommended due to the risk of fouling moorings and the strong run of the tide. There is a Harbour Authority buoy here marked 'ECC Visitors' that can be used for short stays, at £2 for a part day or if waiting to go into Exmouth Dock. This is normally serviced by the useful water taxi that operates from Exmouth Dock to anywhere in the lower Exe, call sign *Conveyance* VHF channels 37 or 80. With sufficient water, a boat capable of taking the bottom comfortably can always anchor 'round the back' of The Point. Here, out of the main stream, and clear of the moorings, you can dry out on Shelly Bank, hard sand, just off the Exe Sailing Club, and visitors from other clubs are welcome to use the showers, clubhouse and bar when open. This large new building was built several years ago after the previous clubhouse on the north side of the dock entrance was spectacularly destroyed when the quayside collapsed.

The main shopping centre of Exmouth is about half a mile from the dock area and all normal facilities can be found. There is fresh water at the Sailing Club, but fuel is not easily obtained except in cans. There are several small boatyards and a chandler in the vicinity of the docks, engine and electronic repairs can be arranged through them or ask the Dockmaster.

The Bight

Exmouth has been a busy holiday resort since the 1750s, but just a short distance across the water the eastern end of the Warren provides a peaceful contrast. Sadly, the deeper anchorage off it is now very restricted by moorings which run parallel to the beach along the curve of the river known as the **Bight**, opposite No 13 buoy. However, shoal draught boats capable of drying out can still creep inside the moorings where the Warren turns back on itself to form Salthouse Lake, and here you will lie in an eddy comfort-

Looking north at around half-tide the channel between the Warren and Bull Hill bank is narrow but deep

ably out of the fast main stream, just grounding or drying out at LW on firm sand, but take care to avoid the oyster and mussel beds on the Cockwood bank further up-stream. In the evenings, and mid-week this delightful spot is very quiet and you will probably have it all to yourself. In the 1930s this end of the Warren was covered with a large number of bungalows and holiday homes, but the winter storms and shifting sands allow no such perma-nence. As a boy, I can remember the last few houses high on the dunes on the seaward side of the beach, and the fascination of seeing their broken remains after they had been undermined and began to collapse. The sea broke right through the **Warren** in the 1960s during a winter storm, but the massive repair works and new groynes on the seaward side have resulted in a steady growth ever since. The western end of the Warren is a popular golf course, but the lonely eastern end is now a 500 acre nature reserve. Here there is nothing more than the low sweep of the beach, and the low dunes behind, a wilderness of lupins and marram grass. At dusk it becomes the silent haunt of the rabbits and birds; just across the water, the cheerful glare of the fairy lights along Exmouth sea-front somehow emphasise the pleasant isolation of the place and there is nothing, except the rumble of the main line trains along the western shore, to disturb its peace.

In complete contrast, if you decide to walk the mile or so along the beach to the village of **Dawlish Warren,** you will find an unsightly sprawl of amusement arcades, gift shops and even a betting shop, catering for the large holi-day chalet villages. There is also a grocery, newsagent, Post Office and pub. Fortunately, the attractions and the long walk tend to concentrate the masses at that end of the Warren, and it is easy to escape back to its real delights.

Starcross

To avoid the encroaching edge of Bull Hill Bank do not steer directly for green No 15 buoy but keep out towards the large mooring buoys in mid-channel. These are pri-marily for commercial vessels but can be used if you are not too worried about your topsides; however, just to the NW there are four yellow Harbour Authority visitors' buoys for up to 40 feet, marked 'ECC Visitors' with white pick up buoys at £4.50 per day inc VAT, call Port of Exeter or Harbour Patrol VHF channel 12 during normal working hours to check availability.

Above the red No 14 buoy and green No 17 a channel branches away to port towards the mass of moorings off Starcross, where there are limited provisions, a Post Office. two banks open mornings only, and a chemist. Starcross Garage can supply diesel and petrol in cans and they have a convenient all-tide pontoon (about 1m MLWS) moored just south of the end of the pier which you can lie on while you go ashore. If you contact the Garage (Tel: 01626 890225) it is often possible to lie here overnight or they might be able to find a spare mooring. The Starcross Fishing and Cruising Club also has a visitors' mooring off the village, although more are often available if members are away cruising, and visitors are welcomed to their fine clubhouse at Regent House just opposite the railway sta-tion or their new premises in the Brunel Tower at the inner end of the pier. Note, however, that the ferry pier is private and only open on the hour when the ferries arrive and a charge is made for landing.

The main channel continues between red No 16 'Shaggles' and green No 19 buoys, with the line of moor-ings to port marking the edge of the drying Shaggles Sand; to the east of the larger moorings on your starboard hand the whole area of Cockle Sand dries completely at LW. Just south of No 19 buoy, and out of the fairway, there is a useful deeper hole known locally as 'Silver Pit' which provides a very useful deep water anchorage.

Lympstone

From the 'Shaggles' can head up for green No 21 buoy. The drying channel to Lympstone, nestling between two red cliffs on the eastern shore branches off just above No 21 buoy and this can provide an interesting diversion, sounding in on a rising tide, for a quick dinghy ride ashore to this pretty, former fishing village, where thatched cottages mingle with more elegant and substantial Georgian houses. A number of local boats have drying moorings off the small harbour but here the bottom is mud, and it will be a messy walk ashore if you stay to dry out.

Back in the main channel, there is *no* No 23 buoy, leave red No 18 to port, and opposite green No 25 buoy, known as 'Powderham perch' the extensive estate behind the railway embankment is Powderham itself, an elegant deer park, and through the trees you can just glimpse Powderham Castle, the seat of the Courteney family since the fourteenth century and home of the present Earl and Countess of Devon. For many years there *was* a perch at this point but as the long spit of sand is continually extending downstream from the next buoy, green No 27, a buoy is obviously easier to move. From here on the channel is used daily by the local sludge vessel *Countess Wear* and is well marked as it turns north-east past the buildings and slipway belonging to Starcross Yacht Club. This has the distinction of being one of the oldest in the country, formed in 1773. The moorings in the pool off the Club mostly dry at LW.

Next, steer for Nutwell Court, the large Georgian house completely set among trees on the eastern shore. Keep to the starboard side of the channel, close to green No 29, No 31, 'Lympstone Perch' and No 33 buoys following the outside curve of the channel, and do not cut across the shallow edge of Powderham Sand to port. During the summer months, as in several of the other West Country rivers, these upper reaches are frequently used by salmon fishermen who shoot long seine nets from small rowing boats, usually on the last of the ebb and first of the flood when they are likely to be encountered anywhere between here and Topsham. The few licences granted for these fisheries are jealously guarded and passed down from father to son, but disease and pollution have seen a dramatic decline from the heady days when up to 99 prime fish were caught in a single shot of the net. Nor, too, has a monster salmon like the one caught by netsman Dick Voysey close to this spot in March 1924 ever been seen again. Weighing a staggering 61¼ pounds, this was no fisherman's tale – it was preserved and can still be seen in Exeter's City Museum.

The large incongruous blocks of flats on the eastern shore are accommodation blocks for the famous Royal Marine Training Camp at Lympstone, so don't be too alarmed if you hear the sound of gunfire along the shore. Beyond the red No 20 'Nob' buoy the channel lies to port of the next unnumbered starboard hand buoy and not the beacon with a green triangular topmark and yellow *Gas Pipeline* sign. This carries North Sea Gas and runs east-west

across the river; anchoring should not be attempted anywhere in the vicinity. Also at this point the tributary of the River Clyst enters the Exe, winding away through the mudbanks to the distant railway bridge on the northern shore beyond which 'Tremletts', builders of high speed powerboats, have premises at Odhams Wharf.

Exeter Ship Canal

The main channel, much narrower at this point, leads close to the western shore, under the high flood embankment past the green No 35 'Ranje' and No 37 'Barrel' buoys and right ahead, the building half hidden in the trees with a tall isolated pine, is Turf Hotel, at the mouth of the Exeter Ship Canal. The entrance lock is to the left of the buildings and the canal, which runs five miles inland to Exeter, has the claim to fame that it is the oldest pound lock canal in England, opened in 1566, after the river passage to Exeter was blocked with a weir by the Countess of Devon in an attempt to force vessels to use her own port at Topsham. The canal was originally entered above Topsham but larger vessels resulted in the extension to Turf in 1827. Although dwindling in importance the Port of Exeter remained active until the late 1960s, with coasters carrying petrol, timber and coal passing incongruously above the reed beds opposite Topsham to create havoc with the holiday traffic on the once notorious Exeter by-pass, when the swing bridge was opened. Today the M5 motorway bridge sweeps across the wide Exe Valley just upstream of Topsham and although this ended the traffic jams, the bridge's 11m clearance also finally sealed the fate of Exeter's waterborne commercial trade.

Since then the canal has seen increasing amenity use for fishing and pleasure boating, and the fascinating trip to Exeter is quite feasible either by dinghy, outboard and a bit of portage, or in boats not restricted by the 11m height limit. Summer visitors are welcomed at reduced rates of £8 per day inc VAT if they want to lie below the motorway bridge in peaceful non-tidal berths at Turf and opposite Topsham, or at £15 per day inc VAT if they wish to proceed all the way to the heart of Exeter. You must stay in the canal for a minimum of two nights but the charges include the cost of locking in and out, transit and a berth at Exeter Canal Basin (where there are shoreside shower and toilet facilities). Providing a fascinating excursion through a wetland conservation area with alongside berthing close to the centre of Exeter for sightseeing and shopping it is well worth the experience. Arrangements to enter the Canal can be made by phone (01392 74306) or by calling *Port of Exeter* on VHF channel 12. There is a surcharge at weekends.

The excellent Exeter Maritime Museum is based at the Canal basin, with exhibits afloat and ashore making up one of the world's largest collections of full size working craft, including junks, dhows, Fijian outriggers, and the steam tug *St Canute*. Created in 1968 by retired Army Major David Goddard this is not just a static display, many of the boats are regularly sailed by members of the International Sailing Craft Association (I S C A) which runs the museum. Open daily throughout the year 1000-1700, it is easily reached by bus from Topsham.

Turf Hotel

Turf Hotel is a favourite local watering hole, and a popular evening trip down river from Topsham. Anchor south-east of the pier on the edge of the channel clear of the local moorings where there is about 1m LAT and a soft mud bottom or pick up the ECC visitors mooring for £4.50 per night inc VAT, and pay at the Hotel. Do not anchor in the approach to the lock as this is used 2 hours either side of HW by the sludge vessel which often sails in the middle of the night from the sewage works further up the canal.

Landing is easy at the steps and pontoon at the end of

Topsham Quay. Trout's all-tide landing pontoon beyond

At Turf, the channel curves away to distant Topsham

the jetty and a lawn leads up to the old slate hung Georgian building, which is open during normal licensing hours with a good choice of bar food. There is a tap where water containers can be filled at the back of the building. The adjoining basin, a tranquil spot, is popular for laying-up and there is invariably an interesting assortment of boats. From here, the tow path along the canal provides a pleasant walk of just over a mile to Topsham Lock where a ferry runs across to the town. During the summer there is also a regular ferry between Turf and Trout's all-tide landing pontoon at Topsham.

Topsham

The final stretch of the river from Turf to Topsham is very shallow at LW, with less than a metre in many places. However, on the flood, a few groundings are no problem in the soft mud, and the channel is not difficult to follow, leaving Nos 39 and 41 green buoys on your starboard hand, 'Ting Tong' No 22 red can to port, and as the channel curves northwards again, leave the perch with triangular topmark, (traditionally known as 'Black Oar'), to starboard and No 24 red can buoy to port. Topsham lies ahead to starboard, and the last channel buoy, green No 45 is in the approach to the main bulk of the moorings which have a clearly defined fairway between them.

Beyond the large modern block of flats that unfortunately dominates the waterfront, Topsham Quay is easily located by the large brick former warehouse, now an antiques market. Visitors can either berth alongside the quay where you will dry out in soft mud or alongside the private pontoon off Trout's Boatyard if space is available, for £7.50 a night, plus VAT. There is 1m at MLWS on the outer pontoon, which makes it best suited for bilge keels or centreboards; deeper draught boats and those with fin

keels can lie alongside Trout's wall where they will sink into the very soft mud. Diesel and water are available alongside and Trouts also stock Calor gas. Petrol can only be obtained in cans from the garage on the main Exeter road. The yard has a mobile crane and engineering /general repairs can be undertaken. During 1995 they hope to install shower and toilet facilities for visitors.

The only possible anchorage off Topsham town is just downstream of the quay in mid-channel, clear of the moorings on the eastern side of the fairway. There is about 1m here at LAT, but the bottom is gravel and the holding not good, particularly on the ebb when the current can run quite fast. Topsham Sailing Club might be able to provide a mooring if members are away cruising. Technically, you will be charged £4 a night inc VAT for lying alongside the Quay which is administered by Exeter City Council, who are slowly upgrading the facilities. Toilets, water, refuse disposal and electricity are available, and also a sailmaker, but unless someone appears to collect your dues the odds are in favour of a free night or two.

'. . . not so small a Town as I find it represented in some Accounts,' wrote an observer in 1754, 'It has not only one pretty long Street to its Kay (where there is a Custom House) and another below it to a fine Strand, the latter adorn'd with diverse handsome Houses, but several good Bye-Streets branching out several ways; and is, in short, a very pleasant, a considerable, and flourishing place, inhabited by many Persons of good Fashion and Politeness, as well as Ship-masters, Ship-builders etc etc . . . Its chief Market, Saturdays, is well supply'd not only with Shambles Meats of all sorts, but Poultry and other Fowls, Butter Cheese and Fruits, and here being Butchers and Fishermen resident,*

Trout's pontoon looking downstream along fairway

Straight Point to Start Point

Looking upstream from Topsham towards HW; the M5 motorway bridge dominates the Exe valley

there's seldom a total Lack of Provisions of either Kind, neither of very good Bread, nor as good Beer, Cyder, Wine, Spirit, Liquors . . . It may be concluded that 'tis no despicable or mean Town . . .'

Two hundred and fifty years on, the same holds good, with just a few additions to the facilities. Topsham is an attractive little town, relatively quiet but once an important Roman and medieval port, and a particularly prosperous place in the mid-seventeenth and early eighteenth centuries when it flourished as a shipbuilding centre. During the Napoleonic wars no less than 27 warships were built here for the Royal Navy by Davy's yards, and in 1850 John Holman opened a new yard with a plan to rejuvenate Topsham as a major port. Wooden vessels up to 600 tons were built, and a dry-dock capable of handling 1,000 ton ships was opened. But within 15 years Holman died and there the dream ended, the silting river finally sealing Topsham's fate. The link with the name remains, however, for John Holman's great grand-daughter Dorothy established a museum of Topsham's history in her house on the Strand which she left to the town on her death. It is now run by volunteers and open on Mondays, Wednesdays and Saturdays, 1400-1700.

Facilities are still good, with all provisions available, including 'Purchalls' grocery store and a 'Co-op', both open on Sundays 'The Foc'sle' a well-stocked chandler and Admiralty Chart agent, branches of Lloyds and Natwest banks, a launderette on the Exeter road, and the traditional Saturday morning market. No shortage of restaurants and bistros, there are no less than ten pubs most of which do food, several cafes and a fish and chip shop. The attractive cathedral City of Exeter is well worth a visit, either by a fifteen minute ride on the regular buses or the train which connects with the main line at Exeter.

Topsham Sailing Club was founded in 1885 and today its very active cruiser fleet is one of the largest in the West Country. Visitors are welcome, and those two cruising essentials, showers and a bar, are available on Wednesday evenings and at weekends. There is a convenient fresh water tap on their slipway, and an all-tide landing pontoon immediately off the Clubhouse.

The Retreat Boatyard lies half a mile beyond Topsham. Nestling almost beneath the motorway bridge, the yard is accessible two hours either side of HW for an average draught boat. Diesel and water are available alongside and they can provide all the usual services, including a 36 ton crane, marine and electronics engineers, a chandler, and Calor and Camping Gaz.

To reach the Retreat, continue upstream through the moorings past the Sailing Club, leaving the old lock entrance on the port hand, and the outfall beacon with green triangular topmark to starboard. Keep a few boat lengths off the shore and pick your way through the large number of local moorings beyond which the river mostly dries at LW leaving a large area of mud and reedbeds, the narrow channel following the reed fringed edge of the playing field on the starboard hand. The next beacon with triangular topmark off a dinghy park is also left to starboard. From here, the channel holds tight to the right hand shore, overlooked by a number of substantial detached modern houses, and there is a large shallow bank to port, so don't be tempted to cut the corner. It is then a straight run past Retreat House, the large white Georgian building, towards the group of moorings and the pontoon off the boatyard. Do not try to anchor as the bottom is foul, but berth alongside the pontoon, rafting if necessary, as this is only used for short stay visitors. If you wish to stay longer, drying moorings are sometimes available off the yard.

PORT GUIDE: RIVER EXE
AREA TELEPHONE CODE: 01392

Harbour Master: Exeter, City Canal Basin (Tel: 74306)
Exmouth Dockmaster: The Docks (Tel: 01395 269314)
VHF: Normal working hours only, channel 12, call sign *Port of Exeter* or *Harbour Patrol*. Exmouth Docks, channel 14, call sign *Exmouth Dock*, Retreat Boatyard channel 37
Emergency services: Lifeboat at Exmouth. Brixham Coastguard
Anchorages: Below No 19 buoy. Turf. Topsham. Various possibilities within river
Mooring/berthing: Exmouth Dock. Harbour Authority visitors moorings in lower Exe and at Turf. Drying alongside quay or Trout's pontoon at Topsham. Drying moorings at Retreat Boatyard. Lock into Exeter Canal
Dinghy landings: The Point, Exmouth. Starcross Pier. Turf. Public slipways at Topsham, ladders on Quay. Topsham SC. Trout's Pontoon
Water taxi: VHF channel 37 or 80 call sign *Conveyance*
Marinas: Exmouth Docks under development 1995
Charges: Exmouth Dock: (per night not including VAT) up to 20ft £4, 20ft-30ft £6, 30ft-40ft £8, 40ft-50ft £12, 50ft-60ft £16. Harbour Authority (per night inc VAT), visitors' moorings £4.50 per night. Exeter Canal (minimum 2 nights), below M5 bridge £8. Exeter Canal basin £15 including locking and transit. Trouts Pontoon, Topsham, £7.50 per night (inc VAT)
Phones: Dock entrance, Exmouth. Starcross. High Street by Church, Topsham
Doctor: Exmouth (Tel: 01395 273001).Topsham (Tel: 874648)
Hospital: Exmouth (Tel: 01395 279684). Exeter (Tel: 411611)
Churches: Exmouth, Topsham, most denominations. C of E, RC, Baptist, Methodist, Gospel
Local weather forecast: None
Fuel: Diesel alongside at Trout's Pontoon and Retreat Boatyard, Topsham. Diesel and Petrol in cans from Starcross Garage, Starcross (Tel: 01626 890225). In cans from WW Pretty's Garage, High Street, Topsham
Gas: Calor and Camping Gaz, Hancock and Wheeler, Exmouth. Wills Garage, Starcross. Trouts and Retreat Boatyards Topsham
Paraffin: Westaways, Fore Street, Topsham
Water: Exe Sailing Club, Exmouth. Turf Hotel. Topsham SC. Trout's Yard and Retreat Boatyard, Topsham, Topsham Quay
Banks/cashpoints: Exmouth: all main banks have cashpoints. Starcross: Lloyds and Natwest, mornings only. Topsham: Lloyds, (has cashpoint) Natwest. All main banks in Exeter have cashpoints
Post Office: Exmouth, Starcross, Topsham
Rubbish: Bins at Exmouth,Trouts Yard Topsham and Topsham Quay
Showers/toilets: Exe Sailing Club, Exmouth. Starcross Fishing & Cruising Club, Starcross. Topsham Sailing Club. Retreat Boatyard, Topsham. Public toilets Exmouth Dock and Topsham Quay
Launderette: Exmouth and Topsham

Provisions: All facilities, Exmouth, EC Weds. Limited shops, Dawlish Warren. Provisions/chemist, Starcross. Good selection shops, Topsham, EC Weds
Chandlers: Peter Dixon, Docks, Exmouth (Tel: 01395 273248). Lavis & Son, Camperdown Terrace, Exmouth (Tel: 01395 263095). Foc'sle, Fore Street, Topsham (Tel: 874105). (Admiralty Chart Agents), Trout's Boatyard, Topsham (Tel: 873044). Retreat Boatyard, Topsham (Tel: 874720)
Repairs: Dixon & Son, Dock Road, Exmouth (Tel: 01395 263063). Lavis & Son, Camperdown Terrace, Exmouth (Tel: 01395 263095). Trout & Son, Topsham (Tel: 873044). Retreat Boatyard, Topsham (Tel: 874720)
Marine engineers: Dixon & Son, Exmouth (Tel: 01395 263063). Wills Garage, Starcross (Tel: 01626 890225). Retreat Boatyard, Topsham (Tel: 874720). Grimshaw Engineering, The Quay, Topsham (Tel: 873539)
Electronic engineers: DS Electrics, Exmouth (Tel: 01395 265550). Retreat Boatyard, Topsham (Tel: 874720). Grimshaw Engineering, Topsham (Tel: 873539)
Sailmakers: Michael McNamara, Camperdown Terrace, Exmouth (Tel: 01395 264907). Sails and Canvas, The Quay, Topsham (Tel: 877527)
Transport: Regular bus and train connections from Exmouth via Lympstone and Topsham to Exeter. Buses and trains to Exeter from Starcross. M5 Motorway at Topsham. Exeter Airport 15 mins
Car hire: Exmouth and Exeter
Yacht clubs: Exe Sailing Club, 'Tornado', Estuary Road, Exmouth, Devon EX8 1 EG (Tel: 01395 264607). Starcross Fishing & Cruising Club, Regent House, Starcross, Exeter, Devon (Tel: 01626 890582) and Brunel Tower, Starcross (Tel: 01626 891996). Starcross Yacht Club, Powderham Point, Starcross, Exeter, Devon (Tel: 01626 890470). Lympstone Sailing Club. Topsham Sailing Club, Hawkins Quay, Ferry Road, Topsham, Exeter, Devon (Tel: 877524)
Eating out: Pubs, restaurants, cafes and fish and chips, Exmouth. Pubs at Starcross. Good selection pubs, restaurants, fish and chips at Topsham
Things to do: Swimming/walking Exmouth and Dawlish Warren. Visits to ancient Cathedral City of Exeter

RIVER TEIGN

Tides: HW Dover −0450. Range: MHWS 4.8M—MHWN 3.6m MLWN 1.9m—MLWS 0.6m.
Spring rates can attain 5 knots in entrance
Charts: BA: 26. Stanford: 12. Imray: Y43
Waypoint: SW end Den Point 50°32′38N. 03°29′98W
Hazards: Shifting bar, dangerous in east or southerly winds and swell, particularly on the ebb. Strong tidal streams in river. Large part of harbour dries. Low bridge across river ½ mile from entrance

---

Just over five miles west of the Exe, and in many ways similar, the River Teign is, unfortunately, of very limited appeal from a visitor's point of view, suffering from two distinct disadvantages. Not only does it have a bar, but, sadly, the wide and drying upper reaches of the estuary are effectively closed to all but small boats by the 3.5m clearance of the road bridge just above the town of **Teignmouth**. A large part of the harbour also dries, space is at a premium, and within the main channel and the entrance the tides run fast, between 4 and 5 knots at Springs. However, in spite of the difficult entrance, a considerable amount of commercial shipping uses the docks, exporting clay, around which the harbour developed during the nineteenth century, and importing pulp and animal foodstuffs, and vessels up to 2,000 tons regularly berth.

It is very much a working port and visitors are left to look after themselves. There is only one official fore and aft visitor's mooring available off the town which can take up to four rafted boats (max 13m LOA) where you will remain afloat, and anchoring in deep water is impossible due to the number of local moorings, and the fairway which must be left clear. However, for a bilge-keeler, or boats able to dry out, the river is a more attractive proposition, as space can usually be found on the 'Salty', the large bank in the centre of the harbour.

The Bar has a justifiably unpleasant reputation and in fresh winds from the south or east, especially on the ebb, it can become a treacherous area of steep breaking seas and surf. In offshore winds, however, and after half-flood, with no likelihood of southerly or easterly weather in the offing, the river is not particularly difficult to enter, and once inside, perfectly sheltered. Although lit, including floodlights on the Ness during the summer, entry is definitely not recommended at night.

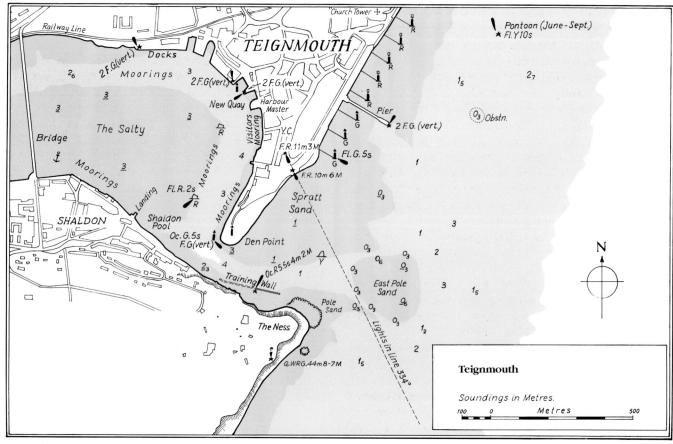

The Teign, with the Ness, left , and Den Point, centre

Approaches

Easily located by the high red sandstone headland of the Ness, topped with trees, on the south side of the entrance, and the conspicuous tower of St Michael's church in the centre of the town on the northern shore, the approach should be made from a position east of the Ness. There is a yellow buoy 1.16M south-east of the Ness, marking the end of a sewer outfall, which is useful to gauge your distance off. Two marks on the Teignmouth seafront, a very tall black pole with a grey stone tower in front of it, provide a transit of 334°T which clears the Ness Rocks. There is a long row of terraced hotels immediately behind the marks, and they lie closest to the second gable of the building. This line leads between the isolated East Pole Sand to seaward and across the tail of the Pole Sand, a spit extending from the Ness. Do not be confused by the small orange buoys with square topmarks in the vicinity - these are used by the local Pilots, and frequently moved. A rocky ledge extends north-eastwards from the Ness, and the shore is littered with large boulders, so don't be tempted to cut the corner but steer towards the conical yellow buoy marking the edge of Spratt Sand, leaving it close on your starboard hand before turning west towards the river entrance. 'Philip Lucette', a white beacon, marks the south side of the channel, standing on a training wall just off the shore which dries 2m; leave it several boat lengths to port, and steer towards the beacon with triangular topmark which marks the end of the low sandy spit, Den Point. **Shaldon,** and a large group of moorings in the 'pool' lie ahead, but the main channel turns sharply to starboard, the moorings and waterfront of **Teignmouth** opening before you.

Here, the tide runs strongly, do not cut the corner, but keep in midstream. The eastern edge of the 'Salty' which dries 3m is marked by three red can buoys which should be left to port, and the fore and aft visitors' mooring lies on your starboard hand; two large red buoys capable of accommodating up to four boats at a time, overnight charge for a 30 footer, £5.87 inc VAT.

The Harbour Office does not maintain a constant VHF watch, and can only usually be contacted when a ship is expected when they use working channel 12, call sign *Teignmouth Port Radio* and would be able to advise as to the visitor's berth availability. New Quay, where the Harbour Office is situated (open weekdays 0900-1230, 1400-1700), lies a bit further upstream on the starboard hand, and it is possible to berth here temporarily if space is available on arrival to contact the Harbour Master, Mr Reg Mathews, (Tel: 01626 773165). If you have decided to anchor and dry out, make your way through the moorings off Shaldon and sound into a suitable spot off the village, clear of the many local moorings. The bottom is sandy shingle, hard, and holding is not particularly good, especially when the tide is running at strength.

A recorded port for nearly 1,000 years, today Teignmouth is rather a mixture. It's a typical beach resort, with reddish sands which were developed early in the 1800s with a traditional pier and elegant seafront, and the usual modern accoutrements novelty golf, a seafront theatre, children's rides and just a few amusement arcades. In marked contrast, behind the town, the jumble of brightly painted fishermen's beach huts, slipways, alleyways, and small quays along the shingly waterfront has its own particular charm. Salmon fishermen work the river in season, often shooting nets right across the river mouth at low water. The famous Morgan Giles Shipyard was located here until it closed in 1969, once employing over 150 local people, designing and building many fine wooden yachts that are still sailing today.

All usual facilities can be found in the town with water on the quay and showers and toilets in Den Point car park, but fuel is only available in cans from the local garage. There is no shortage of shops (early closing Thurs but most remain open in summer), and some open on Sundays, launderette, pubs, restaurants and cafes, Post Office, branches of all main banks, marine engineers (Tel: 01626 872750 or 772324), repairs (Tel: 01626 772324) and electronic engineers (Tel: 01626 776845 or 775960). There are regular main line rail connections.

Shaldon

Shaldon is a much quieter place, an unspoilt waterside village, with a delightful central square surrounding a bowling green. There is less of a choice for facilities, but all the basics are obtainable, including branches of Lloyds and Midland banks, open 1030-1230 Mon-Fri during July and August. There is a general store open daily 0800-2100, 1000-1300 Sundays. There are two chandlers, 'Mariners Weigh' (Tel: 01626 773698) specialising in motorboat gear, and the extremely well-stocked 'Brigantine' (Tel: 01626 772400) on the foreshore, which has just about everything, including Calor and Camping Gaz, but fuel is not obtainable. There are pubs, fish and chips, and several restaurants and tea-rooms. A small ferry runs to Teignmouth if you wish to avoid the row, and for the energetic the coastal footpath up to the summit of the Ness and beyond provides good exercise and fine views. There is also a curious tunnel through the Ness to Ness Cove and, by its entrance, the Shaldon Wildlife Collection.

TORQUAY

Tides:	HW Dover -0500. Range: MHWS 4.9m—MHWN 3.7m MLWN 3.7m—MLWS 0.7m. Tidal streams weak
Charts:	BA: 26. Stanford: 12. Imray: C5. Y43
Waypoint:	Haldon Pier Head 50°27′40N. 03°31′67W
Hazards:	Narrow harbour entrance, often busy. Approach poor in strong SE winds. Inner harbour dries

Centre of the self-styled 'English Riviera' **Torquay** is the nearest thing to Cannes or Nice you are likely to find on a West Country cruise. Once described by Tennyson as the *'loveliest sea village in England'* he would see some considerable changes today from the small fishing village that he knew as a fledgling resort.

Torquay's inhabited existence has a definite pedigree, with the famous underground caves at Kent's Cavern nearby, where spectacular stone age remains dating from *circa* 35,000 BC were found. However, its real growth and prosperity came much later, beginning during the Napoleonic Wars, when Torbay was much used by the Royal Navy as an alternative to Plymouth, being accessible and better sheltered in south and south-westerly weather before the breakwater was built in Plymouth Sound. With the Fleet often anchored in Torbay for months at a time, the town found favour among naval officers, with smart lodging houses rising beside the old thatched medieval cottages. The full transformation to a fashionable resort came with the arrival of the Great Western Railway in 1848, and the unusually mild climate soon established it as the West Country's premier watering place. Large hotels began to rise on the surrounding hills, palms and other sub-tropical plants flourished in the parks and gardens and, as the harbour was enlarged, elegant Georgian terraces began to grow around it.

Torquay developed early as a yachting centre; the Royal Torbay Yacht Club was founded in 1875, and with the completion of the large outer harbour in 1880, the fashionable resort soon became a firm favourite in the grand era of Edwardian yachting popularity. The Regatta was a tradition dating back even earlier, a major event that continues today. Then, with only the drying inner harbour, it was customary for vessels to lie at anchor offshore, and it was here that R T McMullen wrote of a close shave with his 16 ton cutter *Orion* in August 1868, when he joined the fleet of boats anchored in the bay for the regatta.

'. . . Awakened at 2am by the uneasy motion of the vessel, I immediately struck a match over the barometer, and, perceiving that its state was unsatisfactory, hastily dressed myself and called up the men to make snug, as a precaution . . . we first stowed the boat, then housed topmast, and hove the bowsprit short in. Meanwhile the rain was pelting down and the wind gradually backing to the SE, throwing in a nasty sea. The mainsail in very short time became so
thick and heavy with the rain, that the labour of reeving the earrings and taking reefs down was very great indeed. At 4am it blew a heavy gale SE . . . 5am there was a terrible sea, all the yachts were pitching bows under, and most of them beginning to drag home. I was glad to see three or four yachts that lay in our way slip and run for the harbour, although it was only half-tide. Having unshackled our chain at 30 fathoms and buoyed it, we set mainsail with four reefs down, reefed foresail and storm jib and slipped at 5.30am .'

However, they survived the rest of the blow which fortunately veered south-westerly.

Although the outer harbour provided better shelter than

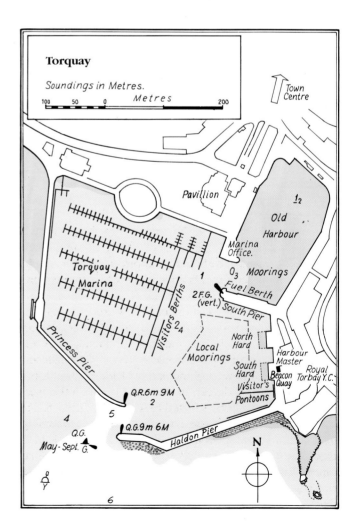

Set in the centre of the 'English Riviera' Torquay has a large marina with excellent facilities

the open roadstead, it still suffered from an uncomfortable surge in south-easterly weather. Amazingly, the short extension to Haldon Pier that eliminated this problem was not built until 106 years later.

A reputation for overcrowding and high harbour dues did much to discourage visitors in the 1960s and 70s, but since the 1985 opening of Torquay Marina with its excellent facilities, Torquay is enjoying renewed popularity.

Approaches

There are no immediate hazards in the approach to Torquay, and there is deep water to the harbour mouth, but as this is narrow, it should be approached with care in strong south-easterly winds which kick up an uncomfortable backwash, making Brixham by far the easier option.

The entrance is only just under 50m wide, and well hidden at a distance because of overlapping end of Haldon Pier. To direct vessels into the harbour mouth between May and September a green conical buoy (QG) is situated approximately 70m west of the end of Haldon Pier – leave this to starboard, and keep to starboard between the pierheads. The yellow buoy situated to seaward marked '5KTS' is one of the limits of the controlled inshore swimming areas around Torbay, within which the speed limit of 5 knots must not be exceeded. Take care in the final approach as the entrance is invariably busy. If closed due to navigational hazards three vertical red balls are displayed on the pierhead – three vertical red lights at night –

and no vessels are allowed to enter or leave.

Entry at night is not difficult, Haldon Pier displaying a quick flashing green light, and Princess Pier a quick flashing red, both of 6 miles visibility, although it is not always easy to pick them out against the blaze of lights from the town. Once inside, two vertical fixed green lights mark the end of the South Pier, the entrance to the inner – old harbour – which is totally given over to local moorings, and dries 1.2m LAT.

The outer harbour has an average depth of 2m LAT, except in the north-eastern corner, east of the fairway leading to the Old Harbour where there are a number of local swinging moorings. Anchoring is prohibited anywhere within the inner and outer harbours and there are heavy penalties for discharging marine toilets within the harbour.

The Marina

The 500 berth marina occupies most of the western side of the outer harbour, and this includes over 60 berths for visitors adjacent to the fairway clearly indicated by a large notice giving berthing directions. Tie up, and report to the marina office. On a par with other marinas, charges can hardly be described as cheap – for a 30 footer it will cost just under £15 per night inc VAT and harbour dues, based on overall length, including bowsprits, davits, etc, and there is a 25% surcharge for multihulls. This includes the use of luxury showers and toilets, fresh water and 24-hour security. Telephones, refuse disposal, a launderette and 240 volt shore power are all available, and a large car park.

Straight Point to Start Point

Entering Torquay leave the green buoy to starboard

Dating from 1912, the extravagant Art Deco Edwardian Pavilion adjoining the marina has been completely refurbished and transformed into an elegant shopping centre, which includes a chandler, cafe/bar, restaurant, delicatessen and off-licence. A 24-hour watch is kept on VHF channel 80, call sign *Torquay Marina.*

For visitors with more austere requirements, the Torquay Harbour Authority can sometimes provide moorings, subject to availability, or berthing alongside the pontoons on the inner end of Haldon Pier. It is best to make VHF contact with the Harbour Authorities on channel 14, call sign *Torquay Harbour,* otherwise, berth, and check immediately with the Harbour Office on Beacon Quay.

Facilities

Facilities in Torquay are excellent, as might be expected in a major holiday resort, and it is a particularly convenient spot for fuelling with diesel, petrol and water all available alongside South Pier from Riviera Boats, who are open 0830-1730 and maintain a listening watch on VHF channel 37, call sign *Torquay Fuel.* There is a large selection of hotels, pubs, restaurants, and although it lacks the restful charm of the peaceful rivers, the bustling contrast can make an interesting change, with plenty of nightlife for the younger members of the crew, although the older hands will probably prefer the more comfortable surroundings of The Royal Torbay Yacht Club. It is open daily and visiting members of other yacht clubs are welcome to use the bar, restaurant, and showers.

PORT GUIDE: TORQUAY
AREA TELEPHONE CODE: 01803

Harbour Master: Captain K.Mowat, Harbour Office, Beacon Quay, Torquay. (Tel: 292429). Office manned Mon-Fri 0900-1300,1400-1700, and weekends also, May-Sep
VHF: Channel 14, call sign *Torquay Harbour,* office hours
Mail drop: Harbour Office. Marina will hold mail and messages for customers. Royal Torbay Yacht Club will hold mail
Emergency services: Lifeboat at Brixham. Brixham Coastguard
Anchorages: In offshore winds, Torquay Roads, 3 cables west of harbour entrance. Anchoring prohibited within harbour
Mooring/ berthing: Harbour Authority pontoon berthing for visitors alongside eastern end of Haldon Pier. Torquay Marina
Dinghy landings: Beacon Quay slipways

Marina: Torquay Marina, Torquay (Tel: 214624). 500 berths, 60+visitors. VHF channel 80 (24 hours), call sign *Torquay Marina*
Charges: Harbour Authority, 25p per foot per day plus VAT. Torquay Marina, £1.40 per metre per day plus VAT
Phones: Beacon Quay. Torquay Marina
Doctor: (Tel: 212489 or 298441)
Dentist: (Tel: 211646)
Hospital: Torbay Hospital (Tel: 614567)
Churches: All denominations
Local weather forecast: At Harbour Office daily. Weatherfax at Marina
Fuel: Petrol and diesel alongside South Pier from Riviera Boats (Tel: 294509) VHF channel 37, call sign *Torquay Fuel*
Water: Alongside South Pier. Torquay Marina. Beacon Quay. Haldon Pier
Gas: Color and Camping Gaz available from Torquay Chandlers – will deliver to Marina customers
Tourist information office: Adjacent to inner harbour slipway
Banks/cashpoints: All main banks in town have cashpoints
Post Office: Centre of town
Rubbish: Compactor, bottle and can bank in Marina car park. Bins on quays
Showers/toilets: Showers and toilets in Marina for customers. Showers available at Royal Torbay YC. Public toilets on quayside. Chemical toilet disposal facility at Marina
Launderette: At Marina and in town
Provisions: Everything available, including basics on Sundays
Chandler: Torquay Chandlers, The Pavilion. (Tel: 211854) Riviera Boats, Beacon Quay. (Tel: 294509)
Repairs: Torquay Harbour Authority has 6 ton crane on South Pier
Marine engineers: Ask at Marina or Harbour Office
Electronic engineers: Ask at Marina or Harbour Office
Sailmakers: Ask at Marina or Harbour Office
Transport: Branch line with regular service to main line at Newton Abbot. M5 motorway 20 mins. Exeter Airport 45 mins
Car hire: Hertz (Tel: 294786)
Taxi: Tel: 213521
Car parking: Many public car parks. Marina car park
Yacht Club: Royal Torbay Yacht Club, Beacon Hill, Torquay TQ1 2BQ (Tel: 292006)
Eating out: Vast selection from fish and chips to top restaurants
Things to do: Torre Abbey-art gallery. Kent's Cavern, underground prehistoric remains. Zoo at Paignton. Aqualand Marine life. Swimming from Torre Abbey sands, 400 yd walk from Marina. Torbay Royal Regatta in August

BRIXHAM

Tide: HW Dover –0505. Range: MHWS 4.9m – MHWN 3.7m MLWN 2.0m – MLWS 0.7m. Tidal streams weak
Charts: BA: 26. Stanford: 12. Imray: Y43
Waypoint: Victoria Breakwater Head 50°24′29N 03°30′70W
Hazards: Busy fishing and pleasure boat harbour. Inner harbour dries

Brixham and Trawlers are synonymous. In the last century the port was famous for its powerful gaff ketches that regularly fished as far afield as the North Sea. Today, a large fleet of sizeable steel vessels are graphic evidence of the remarkable reversal of the port's commercial fortunes; twenty five years ago, the fishing fleet was struggling to survive. This remarkable turn around came with the formation of the Brixham and Torbay Fishermen's Cooperative and, with a united effort, new, much larger vessels were introduced, and the fish quays extended to accommodate them. Since then the port has flourished to become, once again, one of the most prosperous in the country, a success story that is impressive when compared to the recession that has dogged so many other places.

With the expansion of the port of Brixham in recent years much has changed since I first began to visit it thirty years ago when there was far less commercial activity, and always plenty of room to anchor; frequently, we used to nose our way in and find space to dry out in the inner harbour alongside the wall. Today, first and foremost, Brixham is a busy fishing port, secondly it is an important pilotage station with large vessels regularly entering

Brixham Harbour and Marina are of deep and easy access in most normal conditions

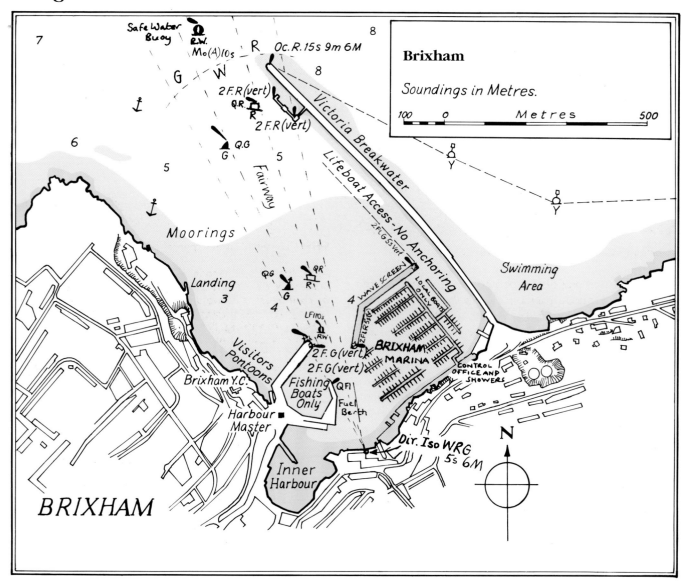

Torbay to take on or land Channel Pilots and there is a busy fleet of trip and angling boats. However, since 1989 when Brixham Marina opened, this major improvement has seen it become an increasingly popular port of call for visitors, of deep and easy access in most conditions.

Approaches

Brixham's half-mile long Victoria breakwater has a white lighthouse at the outer end (Occ R 15 secs) and is easy to locate about a mile west of Berry Head. However, keep clear of the breakwater end as there is often a lot of fishing boat movement in the vicinity and there is an IALA Safe Water buoy, with vertical red and white stripes (Mo (A) 10s) 200m north-west of the breakwater light to mark the beginning of the harbour fairway. Entering, this buoy should be left to port, and vessels should then keep to the starboard side of the 75m wide fairway which is clearly marked by port hand (FlR) and starboard hand (FlG) buoys right up to the fish quays and the Marina. At night the fairway is covered by the white sector of a (Dir Iso RWG 5s) light located on the southern side of the harbour which has a quick flashing light immediately above it to help identify it against the background lights of the town. The

eastern end of the Marina wave screen breakwater is indicated by (2 Fl G 5s vert) lights, the south-western end (2 Fl R 5s vert).

Anchoring is prohibited in the fairway, and the only feasible area is to seaward of the moorings on the west side of the harbour which is sheltered in all but north-westerly winds, and it can also be uncomfortable in north or north-easterly conditions when a ground swell runs in. The bottom is foul in places and a trip line is essential. Anchoring is also prohibited in the fairway along the inside of Victoria Breakwater which is kept clear for the lifeboat. An alternative anchorage, though less convenient for the town, is Fishcombe Cove just west of the harbour, sheltered and peaceful in southerly and westerly weather, and also free.

However, most visitors to Brixham, will probably opt for the convenience of the 484 berth Brixham Marina which now fills the greater part of the south-eastern corner of the harbour, and is well protected by its wave screen breakwater. Up to 100 visitors' berths are usually available, ideally call ahead on VHF Channel 80 call sign *Brixham Marina* for berthing instructions, otherwise approach along the fairway, and once past the western end of the wave screen, either berth anywhere on the outer end of 'D'

Brixham's colourful inner harbour

pontoon or the long events pontoon which extends from the shore to form the southern side of the entrance to the Marina. Once secure, walk along to the Marina Control Office for further instructions. Vessels over 18m LOA should make prior berthing arrangements with the Dockmaster by VHF or telephone.

The facilities are very good, diesel (outer end of 'E' pontoons May-Sept, 0900-2000), water, electricity, showers, toilets, launderette, telephones, 24 hour security, bar, restaurant and chandler. The overnight charge for a 30 footer works out at just over £12.80 inc VAT.

From the marina it is a short walk along the promenade to the centre of town which retains much of its original fishing village atmosphere, climbing in a colourful profusion around the steep hills overlooking the drying inner harbour. This is given over to local moorings, repair work, and trip boat activity. A popular but unsophisticated tourist spot, Brixham has a very distinctive character, alive with pubs, cafes, excellent fish and chips and artists painting on the quay.

The busy main shopping centre, Fore Street, follows the valley leading inland from the harbour, and most facilities can be found. There are branches of all main banks, Lloyds, Barclays and Natwest all have cashpoints. Apart from the pubs, cafes, and fish and chips there are several more up-market restaurants, a couple of bistros.

If you continue around the inner harbour and up the steep hill past the fishmarket you will reach the Brixham Yacht Club, where you are made very welcome, with a bar restaurant, showers and fine views over the harbour. Established in 1937, the Club's burgee includes a crown

and an orange in the hoist which is derived from the historical landing of Prince William of Orange in Brixham in 1688, supported by a Dutch invasion force of nearly 15,000, to restore the throne of England to Protestantism, and be crowned as William III. A statue at the head of the inner harbour commemorates the event.

PORT GUIDE: BRIXHAM
AREA TELEPHONE CODE: 01803

Harbour Master: Captain Mike Wier, The Harbour Office, New Fish Quay, Brixham (Tel: 853321). Office manned seven days a week in season 0800-1800, Mon-Fri only in winter, 0900-1700

VHF: Channel 16, works channel 14 call sign *Brixham Harbour Radio* office hours

Mail drop: Brixham Marina. Brixham Yacht Club

Emergency services: Brixham Lifeboat. Brixham Coastguard

Anchorages: To seaward of moorings on west side of harbour, 4 to 5m LAT. Fishcombe Cove, west of harbour

Mooring/berthing: Brixham Marina.

Dinghy landings: At Yacht Club, and public slipways/steps west and east side of outer harbour

Marina: Brixham Marina, Berry Head Road, Brixham, Devon TQ5 9BW (Tel: 882929). 484 berths, up to 100 visitors. VHF channel 80 (24 hours), call sign *Brixham Marina*

Charges: £1.20 per metre per night plus VAT, short stay £5

Phones: Brixham Marina. At Yacht Club. Public phones by Harbour Office and at head of inner harbour

Doctor: (Tel: 882153)

Hospital: Torbay Hospital, Torquay (Tel: 614567)

Churches: All denominations

Local weather forecast: At Marina and Harbour Office

Fuel: Diesel from fuel berth, Brixham Marina (0900-2000). Petrol in cans from garage or alongside in Torquay

Water: Brixham Marina. Tap at Yacht Club

Gas: Brixam Chandlers, Calor and Gaz

Tourist information office: On old Fish Quay

Banks/cashpoints: Lloyds, Natwest and Barclays have cashpoints. Midland does not

Post Office: New Road, centre of town

Rubbish: Marina, bins and can/bottle bank

Showers/Toilets: In Marina and Yacht Club. Public toilets on New Pier, and Fishcombe car park.

Launderette: At Marina and also Bolton Street, centre of town

Provisions: Everything available. EC Wednesday but many shops open In season

Chandler: Brixham Chandlers, The Marina, Brixham (Tel: 882055). Brixham Yacht Supplies, 72 Middle Street (Tel: 882290). Brixham Net Co Ltd, Pump Street (Tel: 856115)

Repairs: Drying grid in inner harbour by arrangement with Harbour Office

Marine engineers: E G Hubbard & Co (Tel:853327) or ask at Marina

Electronic engineers: Quay Electrics (Tel: 853030) or ask at Marina

Sailmakers: Nearest in Dartmouth

Transport: Bus to Paignton to connect with branch line to main rail line. Ferries to Torquay

Car hire: Ask at Marina

Car parking: Multiple storey car park in centre of town

Yacht club: Brixham Yacht Club, Overgang Road, Brixham TQ5 8AR (Tel: 853332)

Eating out: Good selection from fish and chips to restaurants/bistros

Things to do: Brixham Museum, including Coastguard Museum in centre of town. Aquarium. Walks to Berry Head

Brixham Marina

RIVER DART

Tides: HW Dover –0510. Standard Port. Range: MHWS 4.7m – MHWN 3.5m MLWN 2.1m – MLWS 0.6m. Streams attain over 3 knots in entrance at springs

Charts: BA: 2253. Stanford: 12. Imray: Y47

Waypoint: Castle Ledge Buoy 50°19′95N. 03°33′05W

Hazards: Rocks to SW of Mewstone, Homestone Ledge, Western Blackstone (all unlit). Castle Ledge, Checkstone (lit). Approaches rough in strong southerly weather/ebb tide. Fluky winds in entrance. Ferries between Kingswear and Dartmouth. Chain ferry below Dart Marina. Large part of upper reaches dry

'A *shipman was ther, woning fer by weste:
For aught I woot, he was of Dertemouthe . . .'*

Steeped in a seafaring tradition stretching back way beyond Chaucer's time when the Shipman's 'goodly barge' *Maudelayne* plied her trade from *Gootland to the cape of Finistere, and every cryke in Britayne and in Spayne,* the deep, natural harbour of the **Dart** is, as ever, a must on any cruising itinerary. Often the first port of call after the long haul across Lyme Bay, its steep wooded shores and peaceful upper reaches provide a classic introduction to the area. Epitomising the essential character of West Country cruising, it is a foretaste of the further delights to come.

Approaches

Approaching the river from Berry Head and the east it is not difficult to find with the jagged Mewstone providing a distinctive seamark. From further offshore, to the south and the south-west, however, this island is lost against the shore, and the entrance to the river completely hidden in the folds of the 170m high cliffs. It would still be a notoriously elusive haven if our predecessors had not bothered, in 1864, to build a 25 metre high column with a wide base high above Inner Froward point to the east of the entrance, a fine daymark that is easy to spot. On the west side of the entrance, the row of white coastguard cottages about 100m above Blackstone Point, are very conspicuous against the green hillside. However, once located, there are few other problems. Although the entrance is relatively narrow it is extremely deep with the few hazards close to the shore.

Take care, though, in strong south-east to south-westerly winds and an ebb tide. This can run out at over three knots at Springs, after heavy rain, or strong northerly winds. Confused seas will be found in the immediate approaches, and the high surrounding shore tends to create rapid and baffling shifts of wind, a factor that should be considered if entering under sail alone. Once inside, the shelter is excellent.

From the east, give the Eastern Blackstone and the Mewstone a wide berth as there are a number of drying rocks extending to the west, and hold a course for the conical green 'Castle Ledge' buoy (Fl G 5s) which should be left on your starboard hand. Approaching at night, keep both Start Point (Fl (3) 10s) and Berry Head (Fl (2)15s) visible until Castle Ledge buoy is located. This will bring you into the sectored light on the **Kingswear** shore (Iso RWG

Two castles guard the Dart's deep but narrow entrance

3s) which covers the entrance to seaward, an area known as the 'Range', the red sector just clearing hazards to the west, and the green sector hazards to the east. Keeping in the central white sector you have a safe and easy run in, leaving the 'Checkstone' buoy (Fl (2) R 5s) on your port hand. Passing through the narrows, wait until the white sector of the inner light low down by Bayards Cove (Fl RWG 2s) is open before turning north-west into the harbour, keeping in midstream and taking care to avoid the large unlit mooring buoys in the centre of the channel just beyond Kingswear.

Although straightforward at night, the whole approach should really be timed for daylight, the imposing scenery, St Petrox Church and the two castles guarding the narrow entrance are a sight not to be missed, certainly on a first visit. Kingswear Castle is privately owned, but Dartmouth Castle, to port, dates from 1481, and at one time a 750 foot chain resting on six barges was stretched from here to the opposite shore to protect the port from raiders. It is now administered by English Heritage and open daily during the summer (1400-1830 Sundays), with a small charge. The walk out from the town is well worth the effort with fine views of the entrance and there is a pleasant continuation through woods along the cliffs to Compass Cove, a shingly beach safe for bathing.

In daylight a closer approach can be made from the south-west, keeping close to the 'Homestone', a red can buoy marking a rocky ledge with a least depth of just under 1m LAT to the north-west, and a course can then be laid along the shore, keeping clear of the two Western

SDAPL

Looking inland, the River Dart has plenty of facilities for visitors. Darthaven Marina is off Kingswear on the lower right, Dart Marina is in the centre, with Noss Marina further upstream

Blackstone Rocks off Blackstone Point, and the red Checkstone buoy just south of Dartmouth Castle and St Petrox Church, perched on the western shore. Once past the Castle keep to starboard and follow the eastern shore and the line of moorings below the Royal Dart Yacht Club, a gabled red brick building with a flagstaff and dinghy pontoon. Immediately beyond the Yacht Club frequent passenger and car ferries run between Kingswear and Dartmouth. They have priority at all times and care should be taken to keep clear even if under sail, and always pass astern. A 6 knot speed limit exists throughout the river – do not exceed it.

Don't be too taken aback if you hear a loud whistle and the sudden puffing of a steam train close by; right on the Kingswear shore there is the old railway station, now the terminus for the Paignton to Kingswear Steam Railway, a sight guaranteed to send most overgrown schoolboys into instant nostalgia.

Dartmouth and Kingswear

Stretching before you the large, virtually landlocked harbour is an impressive sight. Dartmouth and Kingswear overlook it from their respective steep hillsides in a colourful profusion of pastels, intermingled with the odd splash of black and white half-timber, and upriver, the imposing

facade of the Britannia Royal Naval College which has dominated the view since 1905 when it replaced the old wooden walled ship *HMS Britannia*. Dartmouth has been the Royal Navy's primary officer training establishment since 1863, a proud association with the town which has, in its time seen several Kings of England, and the heir to the throne among the cadets.

At first glance the harbour is one large mass of moorings, but though crowded in the season, you will always be able to find a berth somewhere in the Dart – even during the very popular Dartmouth Royal Regatta which runs for three days at the end of August. If anything, you are spoilt for choice, but one thing is certain, wherever you go, it will cost you something, for the whole river as far as Totnes is administered by the Dart Harbour and Navigation Authority (DHNA) and dues are payable throughout at a rate of just over12p a foot per day plus VAT, which means a 30 footer will pay about £4.40 per night to anchor, which has always been the cheapest option.

The anchorage is on the Kingswear side of the river between the line of large midstream mooring buoys and Darthaven Marina and the private pile moorings. Do not let go too close to the midstream buoys as their ground chains extend for quite a distance. The holding is good, but the ebb can run at up to two knots at Springs, and it is a good

33

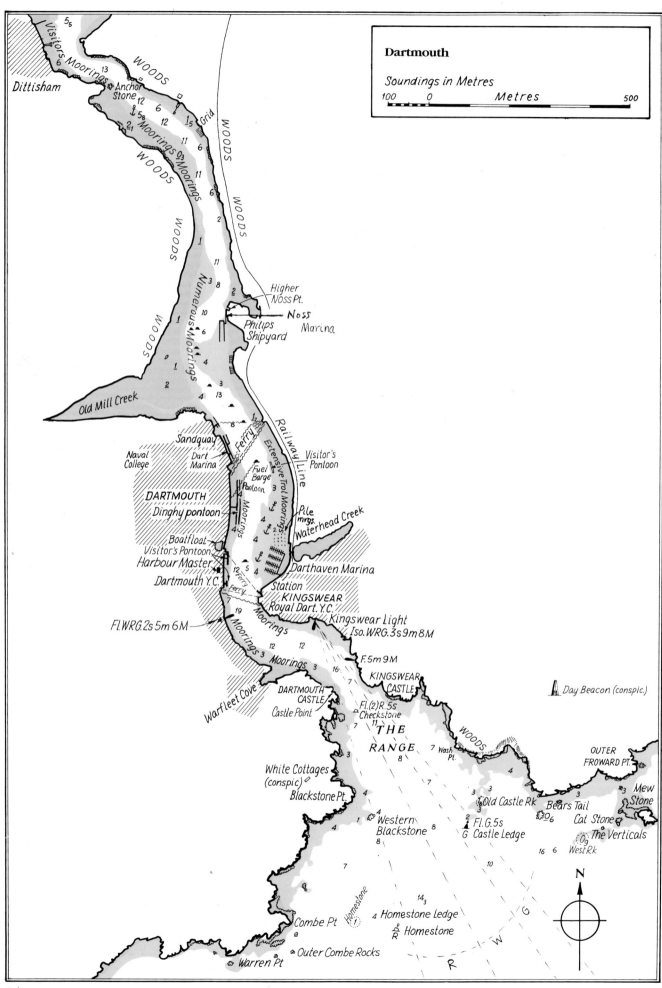

Dartmouth

Soundings in Metres

100   0   *Metres*   500

Dittisham

WOODS

Visitors Moorings

Anchor Stone

Moorings

Moorings

WOODS

WOODS

WOODS

Grid

WOODS

WOODS

Numerous Moorings

Higher Noss Pt.

Noss Marina

Philips Shipyard

Old Mill Creek

Sandquay

Ferry

Naval College

Dart Marina

DARTMOUTH

Dinghy pontoon

Moorings

Fuel Barge Pontoon

Railway Line

Visitor's Pontoon

Extensive Trot Moorings

Pile mrgs.

Waterhead Creek

Boatfloat

Visitor's Pontoon

Harbour Master

Dartmouth Y.C.

Darthaven Marina

Station

KINGSWEAR

Royal Dart. Y.C.

Fl.WRG.2s 5m 6M

Moorings

Kingswear Light

Iso.WRG.3s 9m 8M

Moorings

F.5m 9M

KINGSWEAR CASTLE

Warfleet Cove

DARTMOUTH CASTLE

Castle Point

Fl.(2)R.5s

Checkstone

THE RANGE

Wash Pt.

Day Beacon (conspic.)

WOODS

OUTER FROWARD PT.

White Cottages (conspic.)

Blackstone Pt.

Old Castle Rk

Bears Tail

Mew Stone

Cat Stone

Western Blackstone

Fl.G.5s

G Castle Ledge

The Verticals

West Rk

Combe Pt

Homestone

Homestone Ledge

Homestone

Outer Combe Rocks

Warren Pt

N

R

W

G

Visitors can lie on the inside, lower end of Town Jetty. . .

. . . or on the pontoon islands upstream of Kingswear

row across to Dartmouth. If the anchorage is crowded, wind against tide can also produce quite a few unpredictable antics, and a sudden glut of neighbours in embarrassingly close proximity. The other disadvantage with anchoring is that the Harbour Authority does not allow boats to be left unattended at anchor except for short periods, for shopping or meals. Local fishing vessels work in and out of the moorings and an anchor light is advisable.

There are always moans and groans about harbour dues, but it is probably worth reflecting that the DHNA is a totally independent and self-financing body that maintains the facilities in the river, including the buoyage, and it has been trying in recent years to upgrade the facilities for small craft in the Dart, such as their pontoons on the Dartmouth waterfront. Very convenient for the town, a berth on the lower pontoons (max. 30 feet LOA) works out at nearly £11 per night inc VAT for a 30 footer (double for multihulls). Larger craft can lie further upstream inside the lower end of the Town Jetty; between 1700 and 0900 visitors can also use the outside berths after the passenger boats close down. Here charges are slightly higher, a 30 footer will pay just over £12 per night.

Dartmouth Yacht Club on the embankment extends a friendly welcome to visitors, with excellent showers, a good bar and meals, including breakfast. The Royal Dart YC also welcomes members of recognised clubs, and they, too, offer showers, bar, light lunches and evening meals.

There is a fresh water tap, a very convenient dinghy pontoon and they also have a number of fore and aft moorings, for up to 15 tons, available just downstream for visitors to hire.

The DHNA have a number of cheaper berthing options for visitors, with a pontoon off the northern end of the Embankment for boats up to 26 feet LOA and 4 foot draft, and a long pontoon for larger craft just upstream of the large block of private pile moorings above Kingswear. Both pontoons are free floating without access to the shore. If you do not wish to use your own dinghy there is a convenient water taxi service that can be hailed on VHF channel 16, working channel 06 call sign *Yacht Taxi.* Further upriver, below and off Dittisham they have a number of visitors' moorings, on these or the Kingswear pontoons a 30 footer will pay around £8.80 a night inc VAT. Berths and moorings are normally allocated by the 2-3 River Officers who are on the water during working hours (0900-1700, Sat 0900-1200) and monitor VHF channel 11 and 16 call sign *Dartnav.* If you arrive after hours take whatever seems to be the most convenient option.

Advance booking of moorings is also possible with a non-refundable charge of £10.00 which is deducted from your mooring fees if you take up the reservation. Temporary berthing is also possible alongside the north and south embankment in available space when the tide permits, which is handy for a quick shopping expedition.

Straight Point to Start Point

Dart Marina is the most convenient berth for the town

The north embankment upstream of the ferry pontoon dries on most tides, the south embankment below the pontoon dries at Springs. Quay dues are applicable for anything more than a short stay. Note, too, that there is a very good scrubbing grid on the embankment which can be booked through the DHNA. Both water and electricity are available here and along the Quay.

Finally, there are the three private options, the old established Philip & Son's **Dart Marina** and **Noss Marina** above the higher ferry, and the newer **Darthaven Marina** off Kingswear.

Dart Marina has 100 permanent berths with 15 reserved for visitors but can usually accommodate more if permanent berthholders are away. The nightly charge is 45p per foot, plus VAT and harbour dues, and this includes use of shower, toilet and launderette facilities in their new quayside amenity building. There is a chandler and diesel. Water and Calor Gas are also available and the fuel pontoon is open from 0730, seven days a week. Dart Marina monitors VHF channel 37, call sign *Dart Marina Control,* 0730 - 2000 during the season, 1700 in winter. The Dart Marina Hotel is situated adjacent to the Marina, with a bar and restaurant open to visitors.

The 220 berth Darthaven Marina is cheaper, at 33p per foot per night plus VAT and harbour dues, with showers, toilets, launderette and a large chandler in their new shoreside administration and amenity building. Space for visitors is limited but they can usually fit you in if berth holders are away. Advance booking is not possible, call *Darthaven Marina* VHF channel 80 on arrival or berth in any available space and check immediately with the office. Latecomers will find vacant overnight berth numbers chalked up on the noticeboard by the office, and if your berth is not shown as vacant, move to one that is.

Onshore services include a 30 ton boat hoist, shipwright, engineering and electronic repairs.

Dartmouth

Ashore, Dartmouth has plenty to offer. A busy and popular holiday town, its fortunes were founded around the export of cloth in the middle ages. The magnificent Butterwalk, a half-timbered row of former merchants' houses with their upper storeys perched on granite columns dates from 1640, and its size and fine carvings are an indi-

cation of its former importance. The fascinating Dartmouth Borough Museum is now part of this fine building, and a foray into the steep winding back streets will reveal several other splendid examples of seventeenth century architecture. The attractive cobbled quayside of Bayard's Cove, just beyond the lower ferry, became familiar to many during the 1970s when the BBC used it as one of the locations for its *Onedin Line* series.

A grumpy E E Middleton passed this way in 1869, early on in his 'Cruise of the Kate', an epic single-handed circumnavigation of England in a 21ft yawl. Weatherbound in a south-westerly gale, true to his gentlemanly form, he slept on board and *'went to the Castle Hotel for meals'*. He could still do it today, and in surroundings that have probably not changed a lot, a good meal can be found, overlooking the Boatfloat which is the focal point of the town's waterfront. This enclosed dinghy harbour entered under a bridge is full of local boats but is not the ideal place to leave your dinghy as it dries completely. There are ample landing steps along the embankment – but remember to pull your dinghy clear so that others can get alongside – or berth on the inside of the lower visitors' pontoon. In addition, just below the higher ferry, dinghies may be left on the outer end of the long low tide pontoon off the embankment.

Facilities

Facilities are excellent, with all normal provisions, a good delicatessen, many pubs, good restaurants and cafes and a launderette. There is, however, no branch of Barclays Bank; both Lloyds and Natwest have cashpoint facilities. There are several chandlers, and all repair facilities available, from rigging to divers. Diesel is available from Dart Marina and also the convenient fuel barge moored in mid-river which has some of the cheapest diesel in the West Country and can be contacted on VHF channel 06, call sign *Fuel Barge* It is open 0900-1700 – if reasonable quantities are required after 1700 telephone 834136. Petrol is not available alongside, but can be obtained in cans from Dartmouth Motors, close to the embankment. Water is in plentiful supply from taps on the embankment, all marinas, and the fuel barge. One other facility worth mentioning is the large converted Dutch barge *Res-Nova* moored in the centre of the river just upstream of the fuel barge. This is the home of the Dart Sailing Centre, and open to visiting yachts with berthing available alongside, and facilities include a friendly bar, and showers. There is also a water taxi service to the shore. *Res-Nova* monitors VHF channel 80.

Kingswear

Kingswear, by way of a contrast, is a much quieter place. There is a general store, several pubs and bistros, a launderette and, of course, the Royal Dart YC. The steam trains run regularly to Paignton and back: a great way to keep the kids and father occupied for an afternoon.

'I grudged the delay at Dartmouth', continued Middleton,' *but was recompensed in some measure by the natural beauty of the place, which gains its greatest charm, to my way of thinking, from the entrance to the harbour. I pulled some little distance up the river in the dingby, but the scenery*

'Dominating the view since 1905...' the Britannia Royal Naval College; Darthaven Marina is in the foreground

appeared much tamer, not nearly so beautiful as at the mouth.'

Sadly, the poor chap did not pull quite far enough; the real delights of the Dart are definitely to be found upstream. The name 'Dart' is derived from the Old English word for 'oak' of which there are plenty upstream as the river is navigable for ten miles inland to Totnes; at Springs, coasters up to 1,000 tons and 4m draught still discharge timber just below the town, although less frequently than in recent years. The channel is well-buoyed, and well worth exploring on a rising tide if time permits, and a number of secluded and peaceful anchorages can be found in the upper reaches.

The first section of the river as far as Dittisham is quite straightforward, wide and extremely deep, with over 5m throughout the channel at LAT. The only real hazard is the upper ferry, a large blue and white vessel that runs across the river on wires and berths just downstream of the Dart Marina. Always pass astern and give it a wide berth, as the wires close to it are well up in the water.

Beyond Dart Marina, to port, the jetties and moorings all belong to the BRNC, and a wary eye should be kept out for the numerous small Naval craft running around with Midshipmen under training. The wide mouth of Old Mill Creek dries completely, and the hulk on the northern shore is the remains of one of the last Irish three-masted trading schooners, *Invermore*, abandoned after an abortive attempt to sail her to Australia.

Opposite, the famous shipyard of Philip & Sons at Noss has been building and repairing larger vessels since 1858 and it was also the birthplace of several famous yachts, notably Claud Worth's lovely *Tern IV* and Chay Blyth's first *British Steel*. Fronting the yard, the 190 berth **Noss Marina** was originally restricted to permanent berth holders, but under new re-organisation about 15 berths are now reserved for visitors, more if permanent berth holders are away, providing a quieter alternative to Dart Marina. Here you will pay 35p per foot per night plus VAT and harbour dues; shore facilities are limited to toilets/showers and telephone, but there is a regular launch service to Dart Marina during the day. Noss Marina monitors VHF channel 37 from 0730 to 1800 daily, call sign *Noss Marina*.

From here on, the shores are steep and wooded, the trees brushing the high tide mark, a wide reach which narrows as you approach the **Anchor Stone,** a 3.7m drying rock with a beacon and square orange topmark, which should be given a good berth, and left on your port hand. The tight passage between the rock and the western shore is not to be recommended. As its many victims will confirm, it is a spectacularly embarrassing spot to become stranded!

Dittisham

Sadly, anchoring is no longer allowed off Dittisham because of the extensive moorings and the anchorage just downstream of the Stone is the nearest, if not the most convenient for the village. Sound in just to the edge of Parsons Mud, upstream of the local moorings. The tide can run quite hard here and it can be uncomfortable if the wind is fresh against it; the mud drops away steeply and it is not unknown for boats to drag into the deeper water. A brave row against the ebb, an outboard puts the Ferry Boat Inn within a much more tenable reach! However, if you are merely looking for a peaceful anchorage in sylvan surroundings and wish to go nowhere else you have found it – at least, after the trip boats wind up for the day.

There are a number of DNHA visitors' moorings available off Dittisham, subject to availability, and during the season you will invariably have to raft up. Landing is easy at all states of the tide at the long dinghy pontoon but do not obstruct or leave your dinghy on the end as this is used by the passenger ferry to Greenway Quay, opposite.

Water is available from a public tap by the Ferry Boat Inn which is conveniently situated almost on the foreshore and can provide a good selection of food and drink in convivial surroundings. From here Dittisham, pronounced locally as 'Ditsam' straggles up the steep hill, a mixture of pretty thatched stone and cob cottages leading to the village centre, which has a grocery store, Post Office, phone and another pub, the Red Lion, with fresh local salmon one of its specialities during the season.

There are fine views upriver from higher Dittisham.

The Dart below Dittisham has steep, wooded shores

Widening into a shallower tidal lake, thick with moorings, the water stretches for a mile towards the distant buildings of the boatyard in Galmpton Creek, where the steep wooded shores fall away into gently rolling fields.

Above the moorings, the large area of Flat Owers Bank dries at LW, and the main channel keeps close to the wooded south shore, swinging round across the entrance to Galmpton Creek, where the Dartside Quay and Dolphin Shipyard are both situated at the head of the drying creek. Dartside Quay is a major repair facility with 53 ton and 16 ton hoists and a 6 ton mobile crane. On site services include repairs in wood, GRP and steel, engineering, rigging, electronics, and sailmaker. There is a large area of

Dittisham has visitors' moorings off the Ferry Boat Inn

hard standing. Normally, the yard is accessible 2½ hours either side of HW.

A number of moorings lie along the deep water, but care should be taken to avoid the Eel Rock (dries 1.5m), on the eastern edge of the channel south of the gabled boathouse on Waddeton Point. With sufficient water, there is a short cut to the west of Flat Owers, and from Greenway Quay steer straight for Waddeton boathouse, a course of about 020°T, until abeam of the port hand No 1 Flat Owers buoy, before altering slowly towards the second boathouse, Sandridge, and steering about 310°T. There are a few shallow patches but with a couple of hours of flood you should get across.

From Sandridge boathouse, the channel deepens again and swings back close to Higher Gurrow Point. Steer for the port hand beacon ahead at the entrance to Dittisham Mill Creek, keeping close to Blackness Point, and then working back across the river to the delightfully named Pighole Point on the eastern shore. This wide reach of the river is known locally as the 'Lake of the Dart' as it is impossible to see a way out from the centre at High Water. It is not, however, a deep lake for Middle Back, a large drying bank extends right up its centre. Keep close to the line of moorings along the eastern shore and the beacon with a green triangular topmark which marks the entrance to the drying Stoke Gabriel creek. A channel marked by two port hand beacons with square orange topmarks leads to a dinghy pontoon and a tiny quay where it is possible to land, but beware, do NOT proceed any further than the quay; there is a tidal dam right across the creek, submerged at high tide.

Stoke Gabriel

A picturesque Devon village, with three pubs, a hotel and limited provisions, its claim to fame is the massive yew tree in the church yard, reputedly the oldest in England. A fleet of salmon seine boats lies here in the creek, and during the season, 16 March-16 August, the netsmen can be encountered anywhere in these upper reaches, and should be given a wide berth if fishing.

Bow Creek

The channel bears back to the western shore above Stoke Gabriel, and at the mouth of **Bow Creek** passes between No 2 port hand and No 3 starboard hand buoys, turning sharply to the east and narrowing towards the tiny hamlet of Duncannon. Bow Creek is accessible for shallow draught boats a couple of hours either side of HW, but is probably best explored by dinghy. A convenient anchorage is just under the northern shore by Langham Wood Point, or further into the creek, beyond the low promontory, where you will just ground on a muddy bottom at LW. The channel follows the north shore, then crosses to the south bank, and is marked beyond the entrance by red port hand posts with square topmarks right up to the cluster of buildings at **Tuckenhay**. You are now as deep as you are likely to get by boat into the depths of rural Devon. The shoreline is of overhanging trees, and rough pasture right to the water's edge where cows graze and silently watch as you slip past over the brown, soupy water.

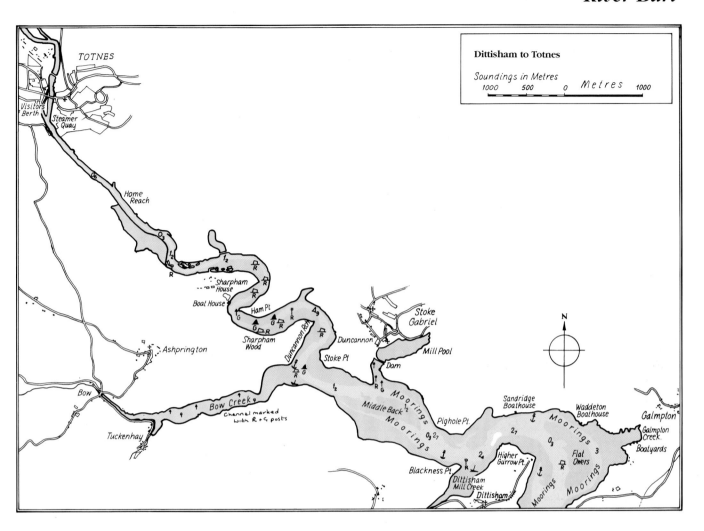

Dittisham to Totnes

Soundings in Metres

1000   500   0   Metres   1000

Tuckenhay is a forgotten, sleepy place, and today it seems amazing that in 1806 it was the ambition of Mr Abraham Tucker to develop it as a major port. His plan never came to much, but the gas house he built did put the village on the map as one of the first places in England to receive gas lighting, and for many years the nearby paper mill produced paper for banknotes. There was, when I first visited the creek, a fine cider factory on the crumbling quayside but developers of a different kind to Abraham Tucker have since moved in and converted the fine stone buildings into luxury homes. The old 'Maltster's Arms', close by, has undergone a transformation too. Now owned by the ebullient TV gastronome Keith Floyd it has been renamed 'Floyd's Inn (sometimes)' and the food, though expensive, lives up to one's expectations. Boats of moderate draught can get up to the quay adjacent to the pub towards HW where it is possible to dry out alongside with the owners permission. Dinner reservations are essential (Tel: 732350), and should you be out of luck, or merely seeking a meal at a more normal price, a ½ mile walk to the head of the creek leads to the 'Waterman's Arms'.

Beyond Duncannon the river remains narrow and, generally, shallow on the inside of the bends. A large number of sizeable pleasure boats run regularly between Totnes and Dartmouth as soon as the tide permits and in these restricted waters passing can be tricky at times. They are plying their trade and you are there for fun, so pull in and let them pass; it is a courtesy they will greatly appreciate.

Totnes

Leaving No 4 buoy to port, keep close to the rocky east shore until the 'Six Knot Speed Limit' sign is abeam, and leave the beacon with red topmark to port, to enter Ham Reach, where the shores are again steep and wooded. Keep can buoys Nos 5 and 7 to port, and Nos 6 and 8 conical buoys to starboard, then remain close to the west shore, under Sharpham Woods, and leave the starboard hand beacon well to starboard as this is inshore of the edge of the bank. Sharpham House, an impressive Georgian mansion standing high above the trees, was reputedly built with 365 panes of glass in its windows, and many other calendar features.

Steer back to the eastern shore beyond Ham Point, leaving the next three can buoys Nos 9, 10 and 11 to port, and into Fleet Mill Reach, avoiding the saltings to port, as the channel lies closer to the starboard bank past the remains of the old paddle steamer *Kingswear Castle* on the shore, then edges back to the western side, passing near to the next port hand beacon. Swinging back starboard, hold a course close to the rocky outcrop on your starboard hand, then turn to port into Home Reach, with the buildings and distant tower of Totnes church in sight at last beyond the flat marshy fields and saltings. The deeper water lies along the centre of the river for the first 200m, then closer to the stone wall to port, then back to the starboard bank up to Baltic Wharf, the timber quay, and passing the boatyard of Totnes Marine Services on Seymour Wharf to port, you will reach Mill Tail, where the river divides. The starboard channel leading to the bridge is full of local moorings, and the Steamer Quay to starboard is in constant use by the pleasure boats. The only berths available for visitors are by the 'Steam Packet Inn' (Tel: 863880) where there is about 20m of quayside by the pub garden with nearly 3m MHWS, and a clean, hard bottom to dry out on. Adjacent to the pub's children's play area there is a further 25m of quay with a muddier bottom. On arrival, check with the pub that you are in a suitable spot. It is owned by the travel writer and broadcaster Simon Cole, who welcomes visiting boats, and no charge is made for an overnight stay. The pub has a restaurant as well as bar meals and even does breakfasts. There is a slot metered electricity supply by the quay, a washing machine and dryer, bathroom facilities and fresh water are all available but charged for.

Totnes is a very attractive and unspoiled medieval town, which rises up a steep main street, to a hilltop surmounted by the remains of the Norman motte and bailey Castle. It is a bustling shopping centre, with all normal facilities available, pubs, restaurants and banks, and is renowned for its secondhand bookshops. There are several supermarkets, including a large 'Safeway' open late and from 1000-1600 on Sundays. Fuel is only available in containers from the local garage.

Finally, on Tuesday mornings throughout the summer, be warned - there is a special market when the locals dress in Elizabethan costumes, so don't be too surprised when your shopkeeper emerges in ruff and pantaloons.

PORT GUIDE: RIVER DART
AREA TELEPHONE CODE: 01803

Harbour Master: Captain C J Moore, Dart Harbour & Navigation Authority, Dart House, Oxford Street, Dartmouth TQ6 9AL (Tel: 832337). Weekdays 0900-1730. Three harbour patrol dories during season

VHF: Channels 16 and 11, call sign *Dartnav*, office hours

Mail drop: Harbour office, Marinas, Royal Dart YC

Emergency services: Lifeboat at Brixham. Brixham Coastguard

Anchorages: Between large ship buoys and moorings on Kingswear side of river. Below Anchor Stone. Various possibilities upriver – ask Harbour patrol. Vessels must not be left unattended at anchor for any length of time

Moorings: Harbour Authority visitors' pontoons on Dartmouth waterfront, upstream of Kingswear and visitors' moorings off Dittisham. Check availability with River Officers. Royal Dart YC fore and aft moorings on application. Berthing alongside Dart Sailing Centre barge *Res-Nova*

Dinghy landings: At pontoons. Steps on embankment and low tide pontoon

Water taxi: VHF channel 16, works 08, call sign *Yacht Taxi*

Marinas: Dart Marina, Sandquay, Dartmouth (Tel: 833351). 15+ visitors' berths available.VHF channel 37, call sign *Dart Marina Control*. Darthaven Marina, Kingswear (Tel: 752545) visitors' berths available.VHF channel 80, call sign *Darthaven Marina*. Noss Marina, Noss (Tel: 833351) 15+ visitors' berths.

Charges: Harbour dues payable throughout river to Totnes. Average charge for 30 footer: £11 – £12 per day inc VAT, alongside. £8.80 per day inc VAT on pontoons; £4.40 per day at anchor. Dart Marina 45p per foot per night plus VAT and harbour dues. Darthaven Marina 33p per foot per night plus VAT and harbour dues. Noss Marina 35p per foot per night plus VAT and harbour dues

Phones: By Boatfloat, opposite the Butterwalk.Marinas and Yacht Clubs

Doctor: (Tel: 832212)

Hospital: Dartmouth and Kingswear Hospital (Tel: 832255)

Churches: All denominations

Local weather forecast: At Marinas

Fuel: Diesel from Dart Marina 0715 – 1715 and River Dart Fuel Barge VHF Channel 06, 0900-1700, out of hours (Tel: 834136) if reasonable quantity required. Petrol in cans from Dartmouth Motors, Mayors Avenue

Paraffin: Dartmouth Motors

Gas: Calor and Gaz, Dart Marina and Darthaven Marina. Gaz only Battarbees Ltd, Lower Street (Tel: 832272)

Water: Free taps on embankment. At both marinas. Free tap at Dittisham, in front of Ferry Boat Inn. Public tap on slipway at Stoke Gabriel. Steam Packet Inn, Totnes, charged by quantity

Tourist information office: In park by Boatfloat

Banks/cashpoints: Lloyds and Natwest have cashpoints. Midland. No Barclays

Post Office: Victoria Road

Rubbish: Bins on embankment and at Marinas. Floating skip off Dittisham. It is a serious offence to throw rubbish into river

Showers/toilets: Available at Royal Dart YC. Dartmouth YC. Dart Marina and Darthaven Marina. Dart Sailing Centre Barge, *Res-nova*. Public toilets on Embankment

Launderette: Dartmouth Launderette, Market Street. Launderettes at Dart and Darthaven Marinas

Provisions: Everything available. EC Wednesday, but most shops open six days a week during season. Shops on Sundays: Mace, Lower Street. Mr Sam's Supermarket, Ivatt Road (on hill out of town—taxi ride) open 0900-2100 seven days a week

Chandler:. Battarbees Ltd, Lower Street (Tel: 832272). Dart Marina (Tel: 833351). Darthaven Marina, Kingswear (Tel: 752545). Harbour Bookshop, Fairfax Place Dartmouth has wide range of nautical books (Tel: 832448)

Repairs: Philip & Son Ltd, Noss Works (Tel: 833351). Creekside Boatyard, Old Mill Creek (Tel: 832649). Dartside Quay, Galmpton, Churston (Tel: 845445). Dolphin Shipyard, Galmpton, Churston (Tel: 01803 842424)

Marine engineers: Ask at marinas, or contact boatyards above

Electronic engineers: Ask at marinas

Sailmakers: Dart Sails, Foss Street (Tel: 832185)

Riggers: Peter Lucas Rigging, Foss Street (Tel: 833094). Atlantic Spars (sparmakers) (Tel: 833322)

Divers: South Hams Divers, Lower Contour Road, Kingswear (Tel: 752581). VHF channel 06

Car Hire: Ask at marinas

Bus/train connections: Main rail line at Totnes, irregular bus connections but regular ferries to Totnes in season. Paignton and Kingswear Steam Railway, Kingswear, connections from Paignton to main rail line at Newton Abbot

Car parking: Off Mayors Avenue behind embankment

Yacht clubs: Royal Dart Yacht Club, Kingswear TQ6 OAB Kingswear (Tel: 752272). Dartmouth Yacht Club, South Embankment, Dartmouth TQ6 9BB (Tel: 832305)

Eating out: Very good selection of restaurants, and good pubs too

Things to do: Dartmouth Borough Museum, Butterwalk. Newcomen Engine House, Mayors Avenue. Dartmouth Castle/St Petrox Church. Friday Market. Steam Railway. Good walks on both sides of harbour entrance

Regatta/special events: Dartmouth Royal Regatta, three days at end of August. South Western Area Old Gaffers Race around end of July

Passages
START POINT TO RAME HEAD

With some romanticism, Frank Carr summed it up.
'West of the Start one begins to have the feeling that one is at last getting to deep seas and blue water; and although the mouth of the Channel is no more from Ushant to Scilly than thirty-nine leagues, it is quite wide enough and deep enough for small craft to be able to be caught out pretty badly . . .'

After the gentle, protected sweep of Start Bay, the rock fringed coast beyond Start Point can often feel exposed, and almost inhospitable, which, in poor weather, it most certainly is. Here, Devon projects its southernmost point far into the English Channel, prey to the weather and the full strength of the tidal streams this combined assault on the coastline is clearly reflected as you move further west to the high cliffs between Bolt Head and Bolt Tail. In the prevailing south-westerlies a long ground swell, and much larger seas will be encountered once clear of Start Point; carrying a fair tide westwards, the odds are certainly in favour of the wind being against it, and this can produce an exaggerated impression of wind and sea conditions.

From Dartmouth, **Salcombe**, six miles west of Start Point, is the most likely destination for most cruising boats – a particularly unspoiled estuary with excellent facilities for yachts. It does, however, have a bar which is extremely dangerous in strong southerly weather, and certain combinations of ground swell and ebb tide. With a least depth

Start Point looking westwards to Prawle Point

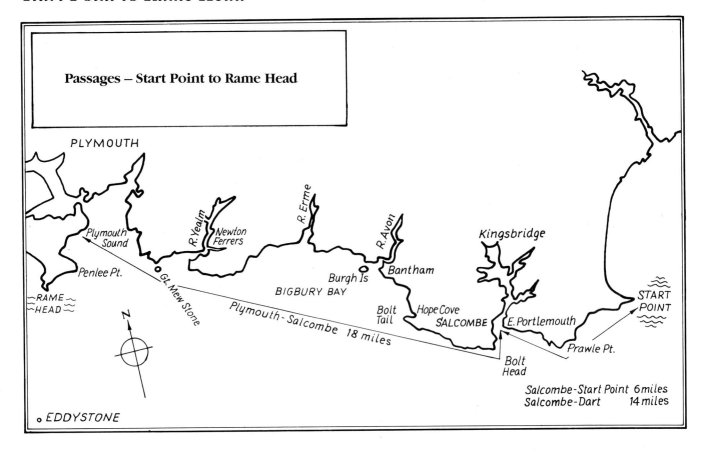

Passages – Start Point to Rame Head

PLYMOUTH

R.Yealm

Newton Ferrers

R.Erme

R.Avon

Kingsbridge

Plymouth Sound

Penlee Pt.

Gt. Mew Stone

Bantham

Burgh Is

BIGBURY BAY

~RAME~ ~HEAD~

N

Plymouth-Salcombe 18 miles

Bolt Tail

Hope Cove
SALCOMBE

E.Portlemouth

START POINT

Prawle Pt.

Bolt Head

Salcombe-Start Point 6 miles
Salcombe-Dart   14 miles

○ EDDYSTONE

of 0.7m LAT it is always best to cross after half flood. **Plymouth**, the next real port of refuge, accessible in all weather, lies almost 25 miles west of Start Point and these factors should be considered before setting out from the Dart, or making a landfall on the South Devon coast.

However, in fair weather, Salcombe presents no problem. Ideally, leave Dartmouth towards local LW which will bring you to the Start with the main ebb running westwards. By the time you reach Salcombe the local inshore flood should ensure ample water over the bar.

Assuming that you pass over a mile to seaward of the Start to avoid the race your course will take you well clear of any inshore dangers such as the isolated Blackstone and Cherrick rocks, and the Sleaden rocks extending about 400m south of Peartree Point. In calm weather, it is possible to pass much closer inshore where the tidal streams turn approximately half an hour earlier, but attain considerably greater strength, up to four knots at Springs. For this reason the fisherman's passage inside the Blackstone is not recommended.

Beyond Start Point the coast is lined by low cliffs, ledges and small rocky bays backed with steep, higher land, a mixture of fields and heathland, with extensive gorse and ferns, and just a few isolated houses. There are no dangers more than three cables offshore but the large numbers of pot buoys can turn the passage into a slalom at times.

Prawle Point, three miles west, is not particularly spectacular. There is a conspicuous terrace of old coastguard cottages immediately east of the Point which has a disused coastguard lookout prominent on its flat, grassy top, with cliffs falling away to a rocky outcrop which forms a natural arch when viewed from the west. As you pass the Point the rusting remains at the foot of the cliffs are of the Greek

registered *Demetrios* wrecked here in December 1992 while on tow to the breakers.

In contrast, to Prawle, **Bolt Head** is a dramatic sight as it emerges in the distance, over 100m high, steep to, and very distinctive, with jagged rocky ridges and pinnacles. The tide runs at up to 2 knots at Springs, and can kick up another small race off Prawle Point which can be uncomfortable at times, and it is best to keep a good half mile offshore.

Emerging beyond Prawle Point, Bolt Head reveals its unmistakeably bold profile

Salcombe lies at the head of the bay formed between Prawle Point and Bolt Head, and a course should be held towards the western side before turning north towards the entrance, which is not easy to spot from a distance. At night, Start Point light is obscured north of a line extending from the Start through Prawle Point; keep the light open across the bay until the leading light into Salcombe is located before turning north.

Off Bolt Head there are two isolated rocks, the

Mewstone (19m) and the Little Mewstone (5m), close to the shore. Overfalls extend to seaward in their vicinity, and this can often be an uncomfortable corner. The high, dramatic cliffs between Bolt Head and Bolt Tail for the next four miles have a particularly rugged and rather grim appearance, an uncompromising lee shore, which should ideally be given a good offing, not only to avoid the dangers, but also the severe squalls that the high cliffs can generate, a phenomenon described with suitable drama by Hilaire Belloc in his 'Cruise of the Nona'.

'As for the spill off Bolt Head, it fell after a clear midnight. . . and it was more than a spill; for it blew for the best part of an hour then ceased. But like its brother off Beachy, it was peculiar to the high land, for it came due northerly whereas the main wind had east in it. I was watching the morning star burning like a sacred furnace on the edges of the black hills when Satan sent that wind and tried to drown three men. But we reefed in time – there were three of us, one for the helm and two to reef; and when dawn broke, and the blessed colours of the east renewed the day, strange! – one end of the boom had three reefs down, and the other only two!'

The main hazards are the Gregory Rocks, least depth just under 2m, half a mile south-east of the Ham Stone, an isolated rock (11m) off Soar Mill Cove, the only major break in the line of the high cliffs, where a grassy valley runs down to the shore. This rock is infamous for sinking one of the last great Finnish grain barques, *Herzogin Cecilie*, when she went ashore in thick fog on 25th April 1936. After seven weeks, the vessel was refloated and towed to Starhole Bay, at the entrance to Salcombe, her rotting and fermenting cargo unwelcome in the harbour. It proved to be her final resting place, for soon afterwards a summer gale from the south-east broke her back, and her remains can still be dimly seen today.

Herzogin Cecilie was wrecked with no loss of life; in contrast the disaster of *HMS Ramillies* in 1760 was catastrophic. A 74 gun ship of the line, set to the east before a severe south-westerly, she mistook Bolt Tail for Rame Head, and realised her mistake too late when steep cliffs, rather than the entrance to Plymouth appeared through the murk. Unable to claw off the lee shore, she anchored, but eventually drove ashore with only 26 survivors from her complement of 734. Few ships have survived once ashore on this particular stretch of coast; it has an uncomfortable feel about it, and I, for one, invariably feel a certain relief once it is safely astern.

The cliffs remain steep right to Bolt Tail, and west of Soar Mill Cove there is a group of three tall radio masts (50m), and then two radio towers (25m) on the grassy plateau along the clifftop. The final hazard just east of Bolt Tail is Greystone Ledge (1.8m) which extends a quarter of a mile to seawards. At Bolt Tail, a precipitous cliff with an isolated rock at its foot, the coast falls suddenly back into **Bigbury Bay**. The cliffs are steep-to, and if followed 200m offshore the small village and harbour of **Hope Cove** will appear, tucked snugly behind Bolt Tail, where it is possible to anchor in the centre of the bay in offshore winds in settled weather.

From here the Devon coastline bears away to the north-

Bolt Tail, with Hope Cove just opening beyond

west, and the large indention of Bigbury Bay, seven miles wide, is usually passed by most cruising boats, as it lies inside the direct course for the Yealm or Plymouth. Although normally a lee shore, in easterly or northerly weather it can be explored further, visiting the little frequented rivers of the Avon and Erme.

About a third of the way across the Bay, off the holiday resort of Bigbury-on-Sea, is the distinctive hump of **Burgh Island** which lies just to the west of the hidden entrance to the drying **River Avon**, bounded on the east by the beach and sand dunes, and a line of low rock strewn cliffs below the golf course at Thurlestone Links. Further east, isolated Thurlestone rock (10m) is a conspicuous feature, rather like a large boat aground on the beach. Accessible near HW in favourable conditions, the Avon is a particularly beautiful small river, and ideal for shoal draught boats capable of drying out, with excellent shelter once inside. There is a temporary anchorage to the east of Burgh Island if waiting for the tide.

The **River Erme**, three miles to the north-west, although attractive, is very open and only really feasible as a daytime anchorage. Wells Rock and several other shoal patches extend ½ mile south of Erme Head, the eastern side of the river mouth. Within Bigbury Bay the tidal streams are considerably weaker, attaining a maximum of 1 knot at Springs on the flood which rotates in an easterly direction, beginning HW Dover +0415, the ebb, weaker, runs to the WNW, beginning HW Dover -0200.

Closing the western shore, Hillsea Point, rounded and grassy with low cliffs has an old coastguard lookout and flagstaff just to the east. Hillsea Rocks and shallow patches lie up to ½ mile offshore. Do not close the land if bound for the Yealm, but hold a course off the shore towards the Great Mewstone, an unmistakeable large pyramid shaped island, until the marks into Wembury Bay can be located, leading into the well-hidden delights of the **River Yealm**. One of the classic West Country havens, this is popular and inevitably crowded in season.

Dominating Wembury Bay, the **Great Mewstone** (59m) is an impressive lump of rock, with sparse sea turf and ferns on the steep slopes that are now just the haunt of the sea birds that give it its name. At one time, however, it was inhabited and the ruin on the eastern side, a single storey circular stone building with an unusual conical roof, was

Start Point to Rame Head

Approaching Plymouth from the east, the Mewstone is a distinctive pyramid. Rame Head is in the far distance

possibly built with the remains of a small medieval chapel that is recorded as being sited there. The last known inhabitants of the Mewstone were Sam Wakeham and his wife, in the 1820s, a rent free domicile in return for protecting the island's rabbits during the off-season for its owners, the Calmady family, and shooting them when required.

Today, the island is owned by the Ministry of Defence, a wildlife conservation area, and subject of a special study it is closed to the public and also lies within the danger area of the *HMS Cambridge* gunnery ranges situated just inshore on Wembury Head. Full details of this are in the Plymouth section, but the normal inshore course into the Yealm from the east is just out of its limits. However, if gunfire is heard and you are intending to pass to seaward of Plymouth, it is worth calling *Wembury Range* on VHF channel 16 to find out what is happening and advise them of your whereabouts.

Shoals extend nearly ½ mile south-west and east from the Mewstone, and the passage between it and the mainland is rock strewn. Beyond it, the huge bay enclosing the large Royal Naval and commercial port of **Plymouth** opens to the north. With the exception of the Shagstone, an isolated rock on the eastern shore marked with an unlit beacon, there are no hazards for small craft, and deep water is carried right into Plymouth Sound, the large outer harbour enclosed by a central breakwater, with an entrance to the east and west. Tidal streams in the approaches to Plymouth are complicated by a clockwise rotation; at the eastern entrance, however, the main flood commences half an hour after HW Dover; the ebb, HW Dover -0530. Streams in the entrances can attain just over a knot at Springs; within the harbour they can be considerably greater in the Narrows.

The approach to Plymouth is easy at night, being well lit, with plenty of water, and the only real problem is likely to be distinguishing the navigation lights from the huge mass of the city lights which can be seen from a very long way off. Care should be taken, too, to keep a careful watch on other shipping, as the harbour and approaches are invariably busy.

Penlee Point on the western side of Plymouth Sound is a wooded and rocky headland with white buildings low down on the point. Draystone shoals extend two cables to the south-east, and are marked by a red can (Fl (2) R 5s). A mile and a half west, Rame Head, the western extremity of Plymouth Sound, is distinctively cone shaped, its grassy slopes climbing to the conspicuous ruins of a chapel on the summit. It is steep-to, and can be approached to within

two cables, although rocks extend from the western side, and in conditions of wind against tide overfalls will be encountered for about half a mile south of the headland.

In good visibility, eight miles south of Rame Head, the thin pencil of the **Eddystone lighthouse** rises incongruously from the sea. A lurking nightmare for ships running into Plymouth, a light on this curious isolated reef was first established in 1698 by an aggrieved shipowner, Henry Winstanley, who had lost two vessels. His distinctly Heath Robinson structure, somewhat akin to a Chinese pagoda, survived a mere five years before it was washed away, taking, by chance, the luckless Winstanley and some workmen with it. In 1709 John Rudyerd erected a wooden lighthouse with considerably more success, but this was destroyed by fire, not water, after 47 years. Four years later, John Smeaton's fine granite column began a vigil that was to last for 120 years, his design setting the pattern for many others, and were it not for the rocks crumbling beneath it, it would probably still stand there today. Instead, when the present lighthouse designed by Sir J N Douglass replaced it in 1882, Smeaton's dismantled tower was re-erected on Plymouth Hoe.

Just awash at HWS, the rocks extend in a radius of about three cables around the lighthouse, a grey tower with a helicopter pad. The major light in this section of coast, it has a 24 mile range, (Fl (2) 10s), and a fixed red sector 112°-129°T covers the paradoxically named Hand Deeps – in reality a shoal with least depths of 7m 3½ miles north-west of the Eddystone, which breaks heavily in bad weather; this and the Eddystone are both very popular with sea anglers and in fine weather there are invariably a number of boats in the vicinity.

In fog, this passage area is not easy, with few aids, and although most of the dangers lie inshore of the 10m line particular care should be taken to avoid being set into Bigbury Bay, the hazards approaching Yealm Head, and the Great Mewstone and surrounding rocks; generally, it is always the best policy to maintain a good offing. Apart from Start Point (Horn 60s), the Eddystone (Horn (3) 60s) and Plymouth West Breakwater end (Bell (1) 15s) there are no other sound signals. As an alternative to feeling your way into Plymouth, where the shipping movements continue in spite of fog, in suitable weather, the anchorage in Cawsand Bay on the western side of the approaches is of easy access, and worth consideration.

The Eddystone light is eight miles south of Rame Head

SALCOMBE

Tides: HW Dover – 0535. Range: MHWS 5.3m – MHWN 4.1m MLWN 2.1m – MLWS 0.7m. Spring ebb attains up to 3 knots off town

Charts: BA: 28. Stanford: 13. Imray: Y48

Waypoint: Wolf Rock Buoy 50°13′49N. 03°46′48W

Hazards: Bar, dangerous in strong southerly weather and ebb tide. Bass Rock and Poundstone (unlit). Blackstone (lit). Harbour very crowded in season. Large part of upper reaches dry

'*S*unset and evening star,
And one clear call for me!
And may there be no moaning of the bar,
When I put out to sea . . .'

So begins Tennyson's famous 'Crossing of the Bar' inspired by the sound of the breaking sea upon Salcombe Bar during a stay in the harbour as a guest aboard Lord Brassey's yacht *Sunbeam* in 1889. Even then, this was a popular, if somewhat exclusive, haunt of the wealthy, who were attracted by the romantic beauty and the beneficially mild, almost Mediterranean climate.

Little has really changed since, except the numbers of visiting boats, and today Salcombe is one of the most popular stopovers for boats cruising the West Country. Away from the main tourist track through Devon, the small town has managed to resist most attempts to over-develop the area, relying totally on its natural attractions. There are several smart hotels, but a limited amount of accommodation ashore, and no amusement arcades or similar diversions to pull in the masses. Apart from the local fishing boats, mostly crabbers, and the Kingsbridge ferry there is no commercial traffic within the river and it is totally committed to pleasure boating in all its forms. As harbours go Salcombe is super-efficient, managing the ever increasingly large numbers of visitors with a polite but necessary authority, for at times, particularly during the regatta weeks at the beginning of August – a tradition dating back to 1857 – the numbers have to be seen to be believed.

As an estuary this is something of an anomaly as there is no Salcombe River flowing out of it, merely a number of small streams from the eight large creeks inland known locally as Lakes. Navigable for four miles to Kingsbridge, a large area of the upper reaches dries, but within the lower half of the estuary there is plenty of deep water, extremely attractive scenery and a number of superb clean sandy beaches. That it never developed as a major port is partly due to its remote position right at the southernmost tip of Devon's South Hams, but also because of the Bar which has a least depth of 0.7m LAT, and becomes very dangerous in strong onshore winds or swell, and ebb tide, when an approach should never be attempted.

Approaches

The approach is memorable, the dramatic steep jagged profile of Bolt Head forming a succession of pinnacles and gullies rising 100m to a flat grassy top. If arriving from the

From the south Salcombe's entrance is dominated by the spectacularly jagged profile of Bolt Head

east, keep a good half mile off the shore to avoid the Chapple Rocks, least depth 2.7m, and close the steep western shore to within 1½ cables of the Eelstone, before turning northwards, but be ready for very strong gusts, and fluky winds beneath the cliffs.

From the west, the Great Mewstone (19m) and Little Mewstone (5m) are large prominent rocks several cables to the south of Bolt Head. They can be passed within a cable and the harbour will then begin to open.

The Bar extends in a south-westerly direction from the rounded fern-covered Limebury Point, footed with low sloping cliffs on the eastern shore, and a leading line is formed by the beacon on the Poundstone, a red and white striped pole with red topmark, and another beacon higher behind it on Sandhill Point, white with a horizontally striped red and white diamond topmark, which displays a Directional Flashing light, (RWG 2s), the white sector 357°-002°T indicating the leading line. It is, perhaps, almost easier to enter at night, as in daytime the beacons are not easy to spot. From a distance Sandhill Point appears as an evenly rounded hill, its wooded slopes dotted with detached houses. Prominent in the centre there is a large red brick house with two white dormer windows in the roof and ivy covered gables at each end – the beacons lie exactly below the left hand gable.

with a little under 2m at LW. ***Because of underwater cables, anchoring is prohibited in the area below and above Mill Bay.*** Just upstream, the next sandy cove on the starboard shore, Smalls Cove, is a popular spot. Sound in to the edge of the beach where there is about 2m at LW. On the opposite side of the channel the first of the pink or yellow visitors' moorings, clearly marked 'Visitor' lie off the large Marine Hotel. In season it is likely that one of the four harbour patrols will meet you in the vicinity and allocate a berth if you have not already contacted them on VHF channels 14 or 12, call sign *Salcombe Harbour*. The overnight charge for a 30 footer on a mooring or pontoon berth works out £9.25 inc VAT and just over £4.60 to anchor.

Just upstream of the Marine Hotel the Salcombe Yacht Club, a large red brick building with a tower and veranda is set up on the hillside; their landing and starting line is on the waterfront below the Club. Here a line of small white buoys with yellow burgee like topmarks run parallel to the shore, marking a fairway that must be used when races are starting, as the centre of the river gets very congested.

The 'Ferry Inn', beside the ferry steps, has a beer garden and bars overlooking the river and is, inevitably, a popular spot for the pundits to gather and watch everyone else's mistakes from behind the safety of a pint.

Beyond the ferry, Batson Lake bears away to port, much

Salcombe entrance: leading marks and Wolf Rock buoy

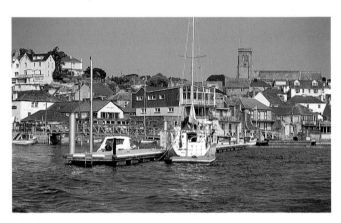

The 'Normandy' short stay visitors' pontoon

Normally, in offshore winds, and preferably after half flood with an absence of ground swell, Salcombe Bar will present no problem.

Once over the Bar, Bass Rock, which dries 0.8m off Splat point can be avoided by steering up to pass close to the conical green 'Wolf Rock' buoy, (Q G), leaving it on your starboard hand. The Blackstone, a large rocky shoal drying 1.5m is marked by a green and white beacon on its western extremity (Q (2) G 8s), and the channel lies between this and the line of red and white beacons on the western shore off the remains of Fort Charles. At night, a pair of quick flashing (Q) leading lights situated just west of Scoble Point, give a transit of 042°T that will take you up the centre of the harbour in clear water with a least depth of 5m right up to the town.

Anchorages and moorings

In settled weather there is a pleasant anchorage just northeast of the Blackstone off the sandy beach at Sunny Cove,

of it drying but there is a clearly marked channel (G/W and R/W poles with topmarks) dredged to 1m LAT leading to the 'Normandy' short stay visitors' pontoon which is linked to the shore by a bridge – vessels over 40 feet and 2m draft should check with the harbour patrol first. This convenient daytime facility is for stays of up to 1 hour only, for shopping or topping up water and must not be used overnight between 1900 and 0700. Just beyond it, the main dinghy landing pontoon is off Whitestrand Quay where the Harbour Office is located. From here, the dredged channel continues along the creek to an all-tide small craft launching slip at Batson.

The main fairway which must be kept clear runs up the centre of the river, and most of the visitors' moorings lie along its edge on both sides; anchoring is not really recommended on the Salcombe side of the channel except just below Snapes Point because of the large number of local moorings. By far the best spot is opposite, off East Portlemouth's sandy beach, upstream of the ferry, towards

Salcombe entrance from above the bar, showing the Wolf Rock buoy, the Blackstone and the beacons off Fort Charles

Ditch End along the edge of the bank which drops off steeply. The tide can run fast , so don't skimp on the cable. Many a relaxed drink at the Ferry has terminated prematurely with the sight of a boat dragging resolutely to seaward!

Salcombe

Historically Salcombe was not always the smart, respectable town that it is today. In 1607 the harbour was a much busier place and the justices reported that the town *'was full of dissolute seafaring men, who murdered each other and buried the corpses in the sands at night'*.

The forlorn remains of Fort Charles at the harbour mouth date from the mid 1600s, the last garrison in England to hold out for Charles I, it endured six rather uncomfortable months while the Roundheads bombarded it from the Portlemouth shore. As a commercial port, the heyday came in the mid-1880s, with a large number of schooners built here to engage in the curious salt cod trade with Newfoundland and Labrador, many from the famous yard of William Date in Kingsbridge. After buying the fish, these small British vessels regularly crossed the Atlantic to sell their cargoes in Spain, Portugal and the Mediterranean and, through this, in turn, a new trade developed bringing citrus and soft fruit to England. Speed was the essence of this financially precarious activity, and it produced a small but extremely fast type of schooner that was able to set large amounts of sail, typical of the ports of South Devon, and Salcombe in particular.

By the beginning of this century, the port had lapsed into total decline, and it was then that its new popularity as a select holiday resort was established. However, apart from the bombardment of Fort Charles, undoubtedly the most dramatic event to engulf the town was the arrival in 1943 of 137 officers and 1,793 men of the US Naval Construction Battalion in preparation for the D Day landings. The Salcombe Hotel was requisitioned as well as many other properties, several rows of old cottages were bulldozed to make a loading ramp which has since become Whitestrand, and the large concrete slipway was built in Mill Bay. A large fleet of landing craft was assembled in the estuary and practice landings carried out on nearby Slapton Ley. On 4 June 1944, 66 vessels sailed from Salcombe; for the small town its invasion had ended, and in Europe, it was about to begin.

Facilities

Nowadays, the invasion is annual, and although not quite as swamped as some of the West Country main holiday resorts, it is probably a good guess that in Salcombe's case 90% of the people walking the narrow streets in mid-summer are doing something with some kind of boat. The town is compact, and rather chic; really just one main shopping street where most normal requirements can be obtained, along with a large number of restaurants, pubs and many bistros.

From a family point of view the attraction of Salcombe has to be the number of beaches. South Sands, at the entrance, is one of the safest and can be reached by a regular ferry, and from here the walk up through the woods

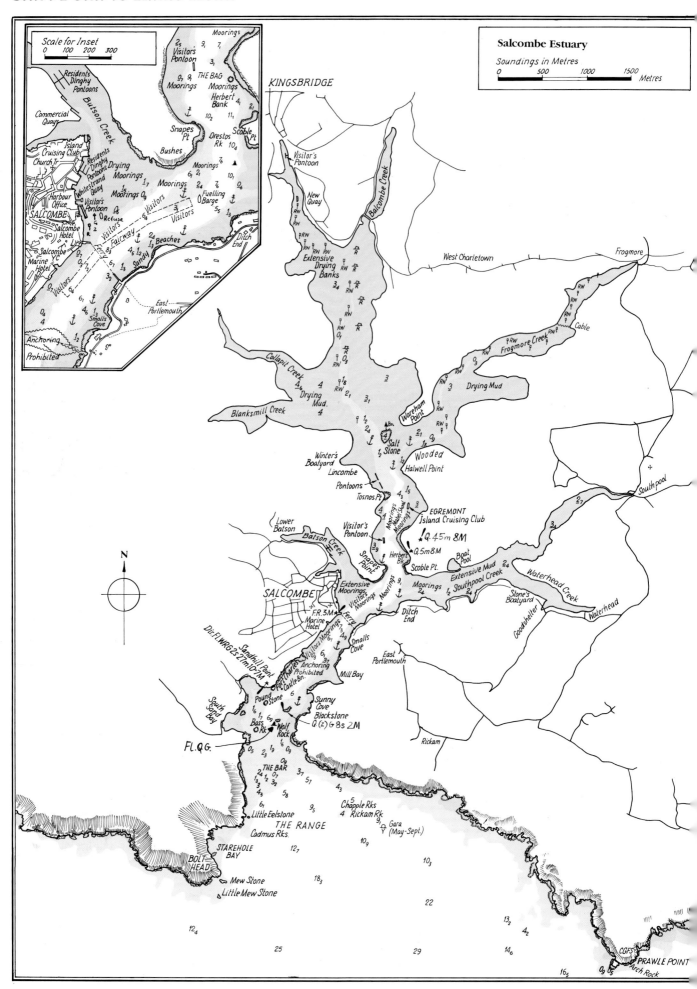

There are visitors' moorings off the town and it is also possible to anchor off the sandy East Portlemouth shore

past the Overbecks Museum and Sharpitor Gardens and on to Bolt Head, all National Trust land, is quite magnificent with spectacular views. If the tide is very, very low, you might just glimpse the ghostly outline of the ill-fated barque, *Herzogin Cecilie*, in her last resting place far below in Starehole Bay.

Greatly favoured by dinghy sailors, the harbour's large area of sheltered, clean water provides excellent sailing, and is one of the reasons that the famous Island Cruising Club has been based here since 1954. An RYA recognised sailing school, its members sail a varied fleet of vessels, from their lovely 72ft 1909 gaff schooner *Hoshi* and the former Brixham trawler *Provident* right down to dinghies. The Club's name was derived from the Island, the area to the south of Shadycombe Creek, 'round the back of the town' where their clubhouse and bar can be found, and in addition, their large accommodation vessel, the former Mersey ferry *Egremont* is moored in the 'Bag' further upriver.

Apart from the ICC clubhouse, here, in this pleasantly salty backwater there lurks all manner of nautical activity, a whole fascinating street of boatyards, engineers, chandleries and other boating businesses, all conveniently close together.

Fuel and water, too, are easily obtained, a fuel barge is moored off the entrance to Southpool Lake, opposite the town, with diesel and petrol, VHF channel 06, call sign *Fuel Barge*, and there are water taps on the short stay visitors' pontoon.

The upper reaches

However, nowhere is absolutely perfect and there are two things that can detract from a visit to Salcombe – the sheer number of boats at the height of the season, and the fact that in a southerly blow, in spite of the apparently landlocked nature of the harbour, a swell, particularly on the ebb, will be experienced as far up as Scoble Point, rendering the anchorage and moorings surprisingly rough and uncomfortable. The solution to both these problems is to seek out one of the quieter, more sheltered spots further up the estuary.

Southpool Lake dries almost to its mouth, but boats able to take the ground might find a spot just upstream of the moorings, grounding at LW. Ideally this is a creek to explore with the dinghy, slipping away from the bustle of Salcombe between the steep wooded shores, following the rising tide as it creeps along the muddy shores. At Gullet Point the creek divides, the starboard arm becoming **Waterhead Lake** where you will find the small, forgotten hamlet of Goodshelter; Southpool Lake is very shallow beyond Gullet but the quiet village of Southpool, where there are no facilities, can be reached a couple of hours either side of HW.

Opposite the entrance to Southpool Lake, Middle Ground, a bank with a least depth of 1.5m, extends south and east from Snapes Point, and the deep water will be found on the Scoble side of the channel. The ebb funnels through this gap, up to two knots at Springs, and opening before you is the traditional upper anchorage of the

Start Point to Rame Head

The popular visitors' pontoon in the 'Bag'

Bag, although now most of it is given over to permanent moorings. A handy anchorage can be found tucked in just north of wooded Snapes Point close to the edge of the drying mud where you will be sheltered from most weather. Alternatively, for the more convivial, just a short distance upstream there is a long visitors' pontoon.

The oldest objection to the Bag used to be the long row to town; the outboard has done away with the worst of that although in a fresh southerly breeze and an inflatable it can still be a long, wet ride. Far more convenient is the taxi service run by the Harbour Authority which can be summoned on VHF channel 12 – charges within the harbour from Blackstone Rock to Scoble Point are 75p a head; from the Bag you'll pay £1.10.

Mabel Shoal inconveniently lies right in the middle of the Bag, with a depth of only 1.2m LAT, and care should also be taken to avoid the rocks extending from Tosnos Point at the northern end of this stretch, the best course to avoid both hazards being to follow the line of moorings along the eastern shore, past the wooded outcrop of Halwell Point where an anchorage can be found close to the shore. Just upstream of Tosnos, the remains on the foreshore are those of the 108ft yacht *Iverna* built by Fay in 1890, and winner of the Big Class 1890-92. When I first visited the Bag in the 1960's she and several other elegant vessels were still afloat as houseboats along this reach.

Nestling in the small inlet at **Lincombe** you will see the buildings and slipways of Winter's Boatyard which is approached by a narrow winding channel marked by small buoys. Ahead, the Saltstone Beacon, a striped pole with green triangular topmark, marks a large drying rock, once used by 17th century Non-Conformists for illegal religious meetings, as the stone belonged to none of the local parishes, and was out of legal jurisdiction. Between it and Halwell Point lies the entrance to **Frogmore Lake**. This, too, dries extensively but a very peaceful anchorage can be found just inside the mouth, with 2m at LW, and with a bit of luck, as evening descends you will find welcome solitude.

After supper, dinghy over for a leg stretch on the lonely pebbly foreshore of Wareham Point, then drift back as dusk descends over the silent wooded shore, the ebb rippling quietly out of the creek, and the peace broken only by the sudden startled cries of the birds settling for the night. Hopefully, the morning will dawn warm and calm, the distant wooded shores indistinct in a gentle, but rising mist as the early sun tries to burn through; it does sometimes happen! If the tide permits, it's time for another trip in the dinghy, following the twisting creek further inland between the rolling fields and grazing cattle to the village of Frogmore. The channel is well marked along its port side by red and white striped poles with square topmarks, and shoal draught boats can get right up to Frogmore on a rising tide, but the final half mile is very shallow. Although not a particularly attractive village, a grocery, Post Office and pub cater for most immediate needs . . .

After the grandeur of the entrance, above Frogmore Lake, the character of the estuary changes completely, becoming wide and flat, meandering gently into the peaceful South Hams, and on a rising tide for an average draught boat it is a pleasant run up to Kingsbridge.

Approaches

Leave the Saltstone well to starboard, and off the wide entrance to Collapit and Blanks Lake the striped red and white poles with red topmarks which clearly mark the western side of the channel should all be left to port.

On the western bank at the north side of **Collapit Lake** there is a conspicuous house with a mooring off it; here the channel divides, the line of red can buoys marking a subsidiary that leads away to the distant group of moorings by the bridge across **Balcombe Lake**. These buoys should all be left well on the starboard hand, and following the poles, the main channel takes you north towards High House Point, where the outskirts of Kingsbridge can be seen, a rather unsightly development of modern houses sprawling across a low rounded hill. Beneath this point the channel turns sharply back to the western shore, there are a number of local moorings along its edges, and upstream **Kingsbridge** begins to appear. The bank on the eastern shore is large and shallow, so do not cut the corner. In contrast to the development opposite, on the western shore trees and fields run down close to the channel, which swings northwards again past a private pontoon with a number of local boats alongside. Just upstream is New Quay, formerly the site of Date's shipyard, and the boatyard and private pontoons all belong to the Kingsbridge ferry which lands here.

Berthing

No public berthing is allowed anywhere along the length of this quay, but keep close, and continue past the ugly modern buildings at its northern end before turning across the entrance of the inlet on your port hand where there are a number of local boats moored. The channel now heads towards the slipway and quay on the western shore which marks the beginning of the tree lined basin enclosing the head of the creek. The visitors' pontoon, with a bridge to the shore, is on the starboard hand; shoal draught boats should berth on the outside, and visitors with deeper draught should use the berths alongside the wall opposite which are clearly marked. Suitable for vessels of up to 12m both berths dry completely to soft mud and are only safely accessible 2½ hours either side of HW, which occurs about 5 mins after HW Salcombe. If intending to stay for a tide, it is advisable to confirm the berthing availability with the Harbour Office before proceeding upstream.

Facilities

Kingsbridge, an unspoilt country town, forms the confluence of four busy roads through South Devon, and although long dead as a seaport, it bustles with activity as an important shopping centre, with a market on Wednesdays. All normal requirements can be found, except fuel; there are many pubs, restaurants and cafes, and after the distinctly nautical atmosphere of Salcombe a taste of the country makes a pleasant contrast, before, once again, returning downstream, towards the sea, and yet another 'Crossing of the Bar'.

For though from out our bourne of Time and Place,
The flood may bear me far,
I hope to see my Pilot face to face,
When I have crost the bar.'

PORT GUIDE: SALCOMBE ESTUARY
AREA TELEPHONE CODE: 01548

Harbour Master: Captain Peter Hodges, Harbour Office, Whitestrand, Salcombe TQ8 8BU (Tel: 843791). Open 0900-1300, 1330-1645 daily during season, otherwise Mon-Fri only. Four Harbour Patrol launches, marked 'Harbour Master' operate in season, 0700-2000, 0600-2200 in peak. 0900-1700 at other times
VHF: Channels 14 and 12, call sign *Salcombe Harbour*, office hours. Harbour patrols, call signs *Salcombe Harbour One, Saltstone, Blackstone* and *Poundstone* all monitor channels 14 and 12
Mail drop: c/o Harbour Office or Salcombe Yacht Club
Emergency services: Lifeboat at Salcombe. Brixham Coastguard
Anchorages: Very restricted space off Salcombe. Off Smalls Cove, E Portlemouth beach, SW and N end of Bag, entrance to Frogmore Lake, W of Saltstone and shoal draft/drying anchorages in many other parts of estuary. Anchoring prohibited in area off Mill Bay
Moorings: 21 visitors' swinging moorings off Salcombe for up to 15m, larger moorings available up to 17m, and three for up to 100 tons TM. Normandy short stay visitors' pontoon off town for up to 1 hour only. Deep water visitors' pontoon in the Bag. Drying visitors' pontoon and berthing alongside wall at Kingsbridge
Dinghy landings: Whitestrand pontoon. Salcombe Yacht Club steps. Ferry steps, E Portlemouth beach

Water taxi: Harbour launches double as taxi service. Call Harbour Taxi on Channel 12, 0900-2315 during season, shorter hours at other times. Charges vary between 75p and £1.10 per person depending on location, landing/departing Whitestrand pontoon. Regular ferries from Salcombe to East Portlemouth, South Sands and Kingsbridge when tide permits
Marinas: None
Charges: Overnight charge for 30 footer, inc VAT, on pontoons or moorings £ 9.25. At anchor £4.62
Phones: Whitestrand Quay
Doctor: (Tel: 842284)
Hospital: South Hams Hospital, Kingsbridge (Tel: 852349)
Churches: C of E, RC
Local weather forecast: At Harbour Office
Fuel: Diesel and petrol alongside fuel barge. Monitors VHF channel 06
Water: On Normandy short stay visitors' pontoon. Tap on Whitestrand Quay
Gas: Calor and Camping Gaz from Salcombe Boatstore
Tourist information office: Town Hall, Salcombe (Tel: 843927). The Quay, Kingsbridge (Tel: 853195)
Banks/cashpoints: Lloyds and Midland only, Fore Street, Salcombe. Cashpoint Lloyds. All main banks in Kingsbridge, cashpoints Lloyds and Midland only.
Post Office: Courtenay Street, Salcombe. Kingsbridge
Rubbish: Floating skip off Whitestrand
Showers/toilets: Salcombe Yacht Club. Public toilets at Whitestrand
Launderette: Fore Street, Salcombe (7 days), also in Kingsbridge
Provisions: All normal requirements in Salcombe. EC Thursday but most shops open during season. All requirements at Kingsbridge
Chandlers: Salcombe Chandlers (Admiralty Chart agents), Fore Street (Tel: 842620). Salcombe Boatstore, Island Street (Tel: 843708). Tideway Boat Construction (Tel: 842987). Sport Nautique, Island Street (Tel: 842130)
Repairs: Tideway Boat Construction, Island Street (Tel: 842987). Winters Marine, Lincombe Boatyard (Tel: 843580). Drying out by arrangement with Harbour Master.
Marine Engineers: Sailing, Island Street (Tel: 842094). Winters Marine, Lincombe Boat Yard (Tel: 843580). Starey Marine Services, Lincombe (Tel: 843655) Reddish Marine (Tel: 844094) Wills Bros, The Embankment, Kingsbridge (Tel: 852424)
Electronic engineers: None
Sailmakers: J Alsop, The Sail Loft, Croft Road (Tel: 843702). J McKillop, The Sail Loft, Ebrington Street, Kingsbridge (Tel: 852343)
Riggers: Devon Rigging, Island Street (Tel: 843195)
Liferaft/inflatable repairs/servicing: Winters Marine (Tel: 843580)
Transport: Regular buses and ferries on tide to Kingsbridge, with connections to main line trains at Plymouth/Totnes via Dartmouth
Car hire: None
Car parking: Large car park at Batson, Salcombe. Large car park on Quay at Kingsbridge
Yacht club: Salcombe Yacht Club, Cliff Road, Salcombe TQ8 8JQ (Tel: 842872)
Eating out: Excellent selection of restaurants/bistros, pubs, cafes and fish and chips
Things to do: Salcombe Museum. Sharpitor Gardens and Overbecks Museum. Spectacular walks on both sides of estuary entrance. Many fine sandy and protected beaches within the harbour
Regatta/special events: Salcombe Town Regatta, first week in August, Salcombe Yacht Club Regatta second week in August. Harbour is also popular for dinghy championships

BIGBURY BAY

River Avon/River Erme:	HW Dover –0523
Charts:	BA: 1613. Stanford: 13. Imray: C6
Hazards, River Avon:	Tidal Entrance with bar, dangerous in onshore wind. Strong currents within entrance. River dries. Murray's Rocks in approach (unlit)
Hazards, River Erme:	Wells Rock to SE, Edward's Rock, East and West Mary's Rocks in entrance (all unlit). Drying tidal river, dangerous in onshore winds

The seven mile stretch of **Bigbury Bay** is, in prevailing south-westerly winds, not a very attractive prospect, and boats heading across it will usually remain a good two to three miles offshore. However, in calm, settled weather, there are several places within it which make an interesting detour from the normal cruising track.

Hope Cove is a convenient anchorage in easterly winds

Hope Cove is a small fishing village tucked away on the north side of Bolt Tail which can sometimes provide a temporary anchorage in easterly winds. There are no off-lying dangers; sound into the centre of the cove between the small pier and the south shore, but beware the Basses Rock closer inshore which dries 1.3m.

The **River Avon** is, however, the real gem of Bigbury Bay, and for owners of shallow draught boats capable of drying out it can be a delightful spot for an overnight stop, providing there is no inkling of a change of weather to the south or west. Although perfectly sheltered within, in an onshore breeze, and any ground swell, the river entrance is inaccessible, and once inside, if the weather does turn, you might not be able to get out for some time . . .

To find it, steer directly for the flattened pyramid of Burgh Island (pronounced 'Burrr') where the beacon marking the drying Murray's Rocks will be seen on the eastern side. Burgh Island itself is tidal, joined by drying sands to the mainland and the trippery resort of Bigbury-on-Sea. At high tide, a remarkable 'sea tractor' with seats on an elevated platform, maintains the link with the shore. In contrast to the mainland resort, privately owned Burgh Island sports an elegant Art Deco hotel, where Agatha Christie wrote several of her books; today, it has become a popu-

lar location for films of similar style and period. On the foreshore, by the landing place, there is also a small pub, the 14th century Pilchard Inn.

The entrance to the Avon is bounded to the east by the extensive sand dunes and beach of the Ham, and apart from a narrow channel, the whole of the approach dries. Ideally it is best to enter an hour before HW (HW Dover - 0523) as the current runs strongly in the entrance. If waiting for the tide, a good sandy anchorage can be found just east of Murray's Rocks in about 2m, which is handy for a dinghy visit to Burgh Island if you wish to pass the time, or to reconnoitre the channel.

The entrance to the river is backed by a high cliff, Mount Folly, and the deepest water over the bar is found by lining up the two conspicuous houses just east of the group of trees on the clifftop. Keep close to the beach on the western side, following the steep shoreline, until the white mark painted on the rock is abeam before turning to starboard towards Lower Cellars Point, which is covered in turf

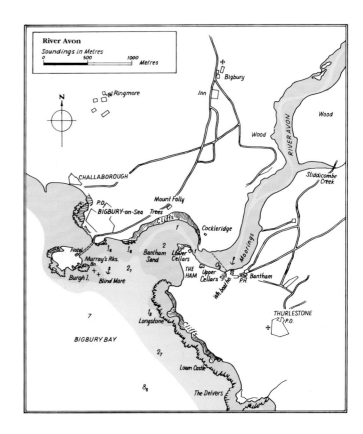

Burgh Island, and the entrance to the River Avon. The sand bar between the island and Bigbury is just uncovering

and ferns, above which protrudes the top of a solitary pine tree. Here the river narrows considerably, with a shingly bank along the northern shore, so steer close to the point, where the current can run fast through the narrows, up to five knots on the ebb if the river is in flood. However, once through the gap, you have entered a real hideaway. The river widens, with the landing beach and houses ahead, and immediately to starboard, there is a sheltered pool off the quaint thatched boathouse and grassy quay, tucked beneath the dense foliage of the steep protective cliff. At first sight, with 2m at LW, this would seem to be the ideal anchorage, but holding is poor, and the current runs strongly. Instead, follow the southern shore upstream towards the white building with a long thatch on the foreshore, and anchor anywhere clear of the few local small boat moorings, where you will ground, and probably dry out at LW.

Peaceful, totally unspoiled, and far removed from the busier anchorages that you have visited so far, this delightful place is a rare find. Disappearing beyond cornfields sloping down to wooded shores, the river winds upstream between drying sandy banks and reedy saltings for another four miles to the village of **Aveton** (pronounced 'Orton') **Gifford**. It is a sleeping, almost forgotten waterway but fifty years ago barges regularly worked their way inland with cargoes of lime, stone and coal for the South Hams farms, and the tiny quays at Bantham recked of the pilchard catches as they were landed and cured. On a rising tide it is well worth further exploration in the dinghy.

Bantham itself is little more than a hamlet, a row of well cared for cob and stone cottages, with deep overhanging thatches, and tiny windows. The village is privately owned, belonging to the Evans Estates, which accounts for the admirable lack of development. There is a village store with basic provisions, Post Office, and the 'Sloop Inn' which does very good bar food. A bigger selection of shops can be found at Aveton where petrol is also available from a garage, if you need an excuse for the dinghy trip.

The large thatched building on the foreshore, in spite of its medieval appearance was actually built in 1937 to celebrate the accession of King George VI, and is nothing more than an enormous boathouse, with two fine figureheads at each end. Within, there is a further surprise, for here you will find Hugh Cater, a boatbuilder who produces magnificent traditional varnished clinker dinghies, and the whole of the upper floor is full of them.

Nearly three miles further north-west, the other forgotten river of Bigbury Bay, the **Erme**, though very attractive, and completely unspoiled, is only really feasible as a daytime anchorage in favourable conditions, and not suitable for an overnight stop. Lacking a sheltering natural breakwater like the Ham, the river mouth is wide open to the south-west, and dries completely beyond the entrance. As it is privately owned by the Flete Estate, and designated as a wild life sanctuary, permission is needed to enter the inner reaches. However, in settled weather and an offshore breeze a pleasant daytime anchorage can be found

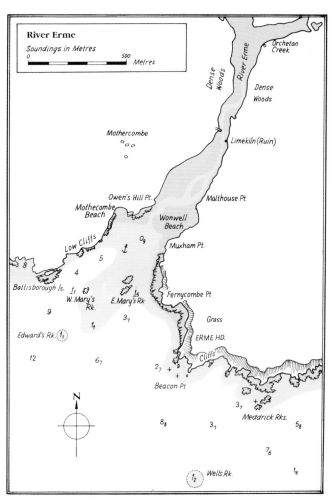

in the mouth off the fine sandy beaches at Mothecombe and Wonwell.

Although there are no offshore hazards, care must be taken in the approach to the Erme, as there are several dangerous rocks, in particular **Wells Rock**, least depth 1.2m LAT just over half a mile south of Erme Head, the eastern flank of the river mouth. Within the entrance there are three more rocks, **West and East Mary's Rocks**, which dry 1.1m and 1.5m respectively and, further to seaward, three cables south of the low grassy Battisborough Island on the northern shore, Edward's rock, least depth 1.1m LAT. Shallow water extends south west from the Mary's rocks, and the deep channel runs parallel to the sloping slab-like cliffs of the northern shore. Keep to the west of all the hazards, and steer for the southern end of Owen Hill, the prominent isolated cliff topped with a group of conspicuous pines. Mothecombe beach lies just to the west, and sounding in towards it, a good sandy anchorage will be found in about 3m, two cables due south of Owen Hill, and inside the Mary's Rocks.

The River Erme is probably unique in that there are absolutely no facilities within a mile; the clean sandy beaches and clear water are perfect for swimming, and if time permits a dinghy trip should certainly be made some of the way upstream, following the tide up the twisting channel between the sandy banks and steep wooded shores. There are few places on the South Coast that remain quite so unspoilt.

The River Avon at half-tide, looking upstream from Bantham

RIVER YEALM

Tides: HW Dover –0540. Range: MHWS 5.4m – MHWN 4.3m MLWN 2.1m – MLWS 0.7m. Attains up to 3 knots in entrance at spring ebb
Charts: BA: 30. Stanford: 13. Imray: C14
Waypoint: Bar Buoy 50°18′56N. 04°04′05W
Hazards: West and East Ebb Rocks, Mouthstone Ledge from east; Outer and Inner Slimer Rocks from west (all unlit). Bar across northern side of entrance (buoyed and lit during season only). Under 2m within entrance at Springs. Not recommended in strong S or SW wind and swell. Newton Creek and large area of upper reaches dry

Undoubtedly one of the classic havens of the entire South Coast, the small, and beautiful **River Yealm** probably owes much of its unspoilt character to the fact that for many years the narrow twisting entrance was a strong deterrent to visitors by sea, making it very difficult to enter under sail alone. With no beaches or attractions for holidaymakers other than the natural beauty of its wooded shores, it was still very undeveloped well into the 1930s, and it is only since the war that the small cottages have become desirable as holiday and retirement homes. Ironically, as reliable auxiliaries overcame the problem of the entrance, they inevitably created another, for the river is now so popular in the height of summer that it is often a tight squeeze to find a berth. In 1989 the Yealm Harbour Authority decided with regret to restrict the number of visitors to a maximum of 90 boats per night at the busiest times of the year, usually during spells of settled weather and particularly on summer Bank Holidays. It is probably best to avoid the river altogether at these times, but they do stress that entry will never be refused to any boat seeking shelter in poor weather.

Don't be put off, though. As long as you are not averse to rafting up, and determined to have a secluded anchorage all to yourself, the Yealm is undoubtedly another essential on any West Country cruise; ideally, visit it early or late in the season, when it is altogether a different place.

Approaches

In spite of the bar, the approach is not as difficult as it might at first seem, although the entrance to the river is totally hidden behind Misery Point. Care must be taken rounding Yealm Head from the east to avoid the Eastern and Western Ebb Rocks, which are just awash LAT three cables south-west of Garra Point. Normally, they can be seen by the seas breaking but are particularly insidious in calm weather. Keep a good half-mile off the shore and do not turn north into Wembury Bay until the conspicuous church tower at Wembury is bearing 010°T.

Approaching from the west, the sloping pyramid of the Great Mewstone provides a fine landmark to locate the river, but keep clear of the Mewstone Ledge, least depth 2m, which extends 1½ cables south-westward and the Outer and Inner Slimers, 2 cables east of the island, which dry 1. 5m and 0. 3m respectively. Continue eastwards until you are on the same 010°T bearing on Wembury Church and these will safely be avoided.

The bottom of Wembury Bay is uneven, and rises quickly to around 6m LAT, a factor that can produce very rough seas in strong winds from the south and west,when, sensibly, no approach to the river should be attempted, particularly with the easy entrance to Plymouth close by.

Holding 010°T, the first indication of the river mouth is usually other boats in the vicinity, and if the wind is off the

Viewed from the west the Great Mewstone conceals the entrance to the Yealm

Wembury church

Yealm entrance. Cellar Bay, right, with leading marks

River Yealm: 'Bar' buoy and leading marks

land, vessels anchored in Cellar Bay. As Mouthstone Point draws abeam the leading marks will be seen above and to the left of Cellar Beach, two white triangular beacons with a black line down the centre, giving a transit of 088.5°T which clears rocky Mouthstone Ledge, extending west for one cable from the point. This line passes across the end of the sand bar that extends from Season Point, which dries 0.6m LAT, and between April and October its southern end is marked by the red can 'Bar' buoy (Fl R 5s) with radar reflector, which should be left on your port hand. This is the only light in the river, and without local knowledge, entry at night, except in ideal conditions, with a good moon, is not recommended, certainly never for a first visit.

Almost opposite, perched on a rock, a green beacon with triangular topmark on a white backboard marks the southern side of the channel, which is about 40m wide and between the two there is a least depth of almost 3m LAT. Abeam of the beacon, turn on to 047°T, towards the square white leading mark with red vertical stripe, high in the gorse on the opposite cliff, unless you are intending to stop off Cellar Beach, which is usually a crowded daytime anchorage if the weather is fair. Across this section of the river on the lowest tides there will not be more that 1.2m until Misery Point is rounded, a sharp turn to starboard into the first inner reach. If in doubt about the water, do not attempt the entrance until a good hour after LW.

Anchorages and moorings

Once inside Misery Point you enter a different world as the sea vanishes astern. The channel deepens to well over 2.5m, and runs between steep wooded banks, with little indication that there is any kind of settlement, except the moorings, among them the first and largest visitors' buoy, capable of taking up to 25m, just north of Misery Point. The second set of moorings lie on the south side of the channel opposite the red can 'Spit' buoy which marks the drying sandy spit extending from Warren Point. In spite of the buoy it's a regular spot for grounding! These two 25 ton moorings can accommodate up to three visitors at a time, and, a short distance further upstream, on the starboard hand, the 150ft visitors' pontoon can accommodate up to about 25 rafted boats at a time.

You are now in the Pool, a fine, landlocked natural haven, surrounded by high hills, wooded right to the water's edge, the traditional anchorage in the Yealm but now very restricted by moorings. Overlooked by the prominent Yealm Hotel, here the river branches, with the Newton Arm leading off to starboard. This dries almost completely, but two hours either side of HW, the villages of **Newton Ferrers**, half a mile upstream on the northern shore and the delightfully named **Noss Mayo**, opposite, can be reached with an average draught. At the head of the creek, another quarter of a mile or so, there is a public quay at Bridgend and boats drawing up to 2m can reach it at Springs, a couple of hours either side of HW, with berths alongside for scrubbing or repairs.

This and the other drying berths alongside Pope's Quay, Noss Mayo, and the scrubbing posts at Clitters Beach, opposite Madge Point, are all available by arrangement with the Harbour Master. Bridgend Boat Co

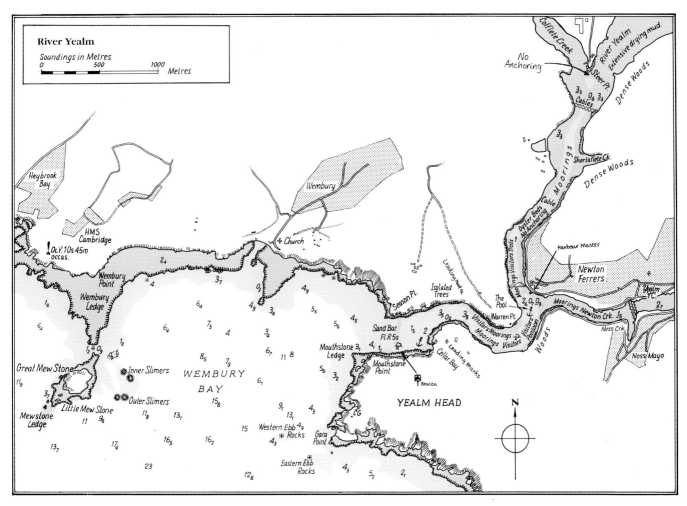

have their premises beside the quay, with a slipway and repair facilities.

Bilge keelers can dry out comfortably on the foreshore off Newton or Noss on reasonably hard ground although the centre of the creek is soft and muddy; take care, however, to avoid the underwater power cables between Newton Ferrers and the eastern side of Noss Creek, which are clearly marked by beacons.

For deep draught boats, the Pool is no longer an easy place to let go during the season as swinging room is so restricted and it is usually necessary to lie to two anchors. As the charge for using a harbour mooring facility is only £1 a night more than lying at anchor, most people opt for the former.

Although well protected from any seas, the river can be surprisingly violent in a blow, as the wind eddying round the high shores creates williwaws, and very strong gusts. On several occasions I have seen small craft laid over almost on their beam ends, and great columns of spume and spray whipped off the water in this seemingly perfect shelter. The Harbour is leased from the Crown Estate Commissioners and administered by the River Yealm Harbour Authority, a statutory non-profit-making organisation who have had to cope with recent large increases in the rent levied by the Crown Commisioners and increase their charges accordingly. Above Warren Point and close to the Wembury side of the river there is a very convenient line of fore and aft trots which can take up to five boats abreast. In addition to these Harbour Authority moorings,

there is also an unwritten understanding in the Yealm where space is at such a premium that visitors can pick up any other vacant residents' mooring providing there is no tender or note attached to it.

The Harbour Master is invariably out on the water first thing in the morning, and after 1640 in the afternoon to assist with settling people in. If you pick up a vacant mooring during the day nip ashore to his office in the driveway of the Yealm Hotel to check on its availablity. There is a large dinghy landing pontoon at Yealm Steps below the Hotel – do not obstruct the outer end as this all-tide berth is kept clear for boats to come alongside for water.

Looking upstream towards Yealm Point and the Pool. 'Spit' buoy, left, and visitors' pontoon on distant right

Facilities

From the Pool, Newton Ferrers is easily reached by dinghy if the tide permits or a 10 minute walk along the path overlooking the steep sided creek. Alternatively, there are two landing places on the south side of the entrance to Newton Arm, opposite the visitors' pontoon, Wide Slip and Parish Steps, and a delightful footpath leads through the National Trust woodlands to Noss Mayo. Newton was presented to Henri de Ferrieres by William the Conquerer as a reward for his assistance in the Norman invasion, which explains the strange name; Noss derives from the Old English for 'promontory'. Both villages have been very well preserved, although in recent years there has been a regrettable spread of modern bungalows and houses around them. Looking out over the creek, the older cottages are immaculately kept, the whitewash looking almost new and doorways overhung with wisteria and roses. Ablaze with flowers their well-tended gardens, neat stone walls and lawns drop away right to the water's edge. There is a Post Office and grocery in both villages; Newton Ferrers also has a butcher, chemist, and Calor/Gaz from 'Village Fayre' up the hill going out of the village.

The Yealm remains a quiet, relaxing sort of place. The walks in the surrounding woods, and beside the river are lovely, reflecting the origins of the river's name which is Celtic for 'kind'. The walk over Warren Point out to Season Point, overlooking the river mouth is particularly fine, with good blackberrying in late summer, it is a memorable spot to watch the sunset over Wembury Bay and the Mewstone. If the timing clashes with other pursuits, there are three pubs to choose from in the villages, the 'Swan' and 'Old Ship' at Noss, with the 'Dolphin' and also the bar at the Yealm Yacht Club on the foreshore at Newton Ferrers. Here, visitors are welcome, and facilities include showers, bar and meals available on Friday and Saturday evenings, and Sunday lunch.

Beyond the Pool, the upper reaches are well worth a trip on the tide, and in a small boat they are navigable for another two miles. Until quite recently the whole of this area was given over to extensive oyster beds but now these only extend for just over 300m upstream of Madge Point. Within this area, marked by notices on posts along the shore, anchoring is strictly prohibited, but upstream the river has been changed considerably by the increase in the number of moorings. The deepest water lies in the centre of the river and depths vary with as little as 0.3m LAT in places and beyond the oyster beds, upstream of the marked underwater power cable, the river is all privately owned by the Kitley Estate. Here, too, anchoring is prohibited throughout as the owner is attempting to maintain it as a nature reserve. At Steer Point the river divides, Cofflete Arm to port and the larger Yealmpton Arm to starboard. Both are muddy and dry at LW, but can be explored by dinghy.

PORT GUIDE: THE YEALM
AREA TELEPHONE CODE: 01752

Harbour Master: Mr M J Simpson, Harbour Office, Yealm Hotel Drive, Newton Ferrers PL8 1 BL (Tel: 872533). 1000-1200, 14301540, daily in season. Afloat in launch early morning and evenings

VHF: No radio watch

Mail drop: c/o Harbour Office

Emergency services: Lifeboat at Plymouth. Brixham Coastguard

Anchorages: Cellar Bay in offshore winds. In the Pool, but very restricted. Drying out in Newton Arm

Mooring/berthing: Harbour Authority has several swinging moorings, to accommodate visitors from 25m – 7m. Visitors' pontoon for up to 25 boats, and fore and aft trots. Local moorings available to visitors if empty. Scrubbing/repair berths available alongside by arrangement with Harbour Master

Dinghy landings: All tide pontoon at Yealm steps, below Hotel. Parish steps, opposite visitors' pontoon. High water landing at Newton Ferrers slip, Noss Creek, Yealm Yacht Club, Bridgend Quay

Water taxi: None

Marina: None

Charges: Harbour dues, per night, inclusive of VAT, up to 20 ft £5. 21-25 ft - £6. 26-30ft - £7. 31-36ft - £8. 37-44ft - £9. Over 44ft - £10. Vessels at anchor £1 per night less; surcharge of £2 per night for multihulls over 30 ft. Cheaper rates for longer periods and boats can be left by arrangement except during July and August

Phones: Outside Post Offices, Newton Ferrers and Noss Mayo

Doctor: (Tel: 880392)

Hospital: Nearest, Derriford, Plymouth, 10 miles. (Tel: 777111)

Churches: Two; C of E, Newton Ferrers and Noss Mayo

Local Weather Forecast: None

Fuel: None

Gas: Calor and Camping Gaz from Village Fayre, Newton Hill. (Tel: 872491)

Water: Hose at outer end of Yealm steps pontoon at all states of tide

Tourist information: None

Banks/cashpoint: None. Post Office will cash cheques for charge

Post Office: Newton Ferrers and Noss Mayo

Rubbish: Bins at Yealm Steps

Showers/toilets: Yealm Yacht Club

Launderette: None

Provisions: All basics available, including chemist. EC Thursday. Groceries available in both Newton Ferrers and Noss Mayo 1000 -1200 Sundays

Chandler: None

Repairs: Bridgend Boat Co, Bridgend Quay (Tel: 872162)

Marine Engineer: J Hockaday (Tel: 872369)

Electronic engineers: None

Sailmakers: A Hooper (Tel: 830411)

Riggers: None

Transport: Six buses a day, weekdays to Plymouth, for main line train connections. Airport at Plymouth

Car hire/ Taxi: None

Car parking: Very restricted

Yacht Club: Yealm Yacht Club, Riverside, Newton Ferrers , Plymouth (Tel: 872291)

Eating out: Meals at hotel in Newton Ferrers, and pubs at Noss Mayo and Newton Ferrers. Meals at Yacht Club, weekends

Things to do: Good walks. Swimming at Cellar Beach

PLYMOUTH

Tides: HW Dover –0540
Range: MHWS 5.5m – MHWN 4.4m MLWN 2.2m – MLWS 0.8m. Ebb streams can attain in excess of 3 knots in Narrows, and 5 knots in upper reaches of Tamar
Charts: BA: 30, 1967, 1901, 1902, 871. Stanford: 13. Imray: C14
Waypoints: Knap buoy 50°19′52N 04°09′94W. East Tinker Buoy 50°19′17N 04°0′23W
Hazards: Shagstone (unlit). Much commercial and Naval shipping. *HMS Cambridge* gunnery range to seaward of Wembury Head. Strong tidal streams in places. Much of upper reaches dry

'Plymouth is a Naval and commercial port and has one of the finest natural harbours in the country. It is not frequently used by yachts. . .'

When my predecessor, D J Pooley was writing the original West Country Rivers in 1957, this paradox held true, and remained so for many years afterwards. Perfect for fleets of warships, the sheer physical scale of Plymouth had little to offer smaller craft, and apart from a few rather dirty commercial basins, there was never a really convenient place for visitors to lie. Most of the better anchorages are a long way from the town, and exposed in bad weather. Today, however, there are four marinas with excellent facilities for visitors. Although still primarily a Naval port, with considerable commercial traffic, including regular ferries to Brittany and Spain, and a busy fishing fleet, Plymouth has also come into its own as a yachting centre and the venue for a number of major events, including the Single and Double Handed Transatlantic races, the Round Britain and the finishing point for the Fastnet.

Approaches

Certainly, there are no natural problems in the approaches to this vast harbour bounded on the east by the Great Mewstone and Rame Head to the west, with ample deep water for small craft throughout. The hazards are entirely man-made, with shipping the main consideration, and much Naval activity in and around the port. Under the jurisdiction of the Queen's Harbour Master, all movements are controlled by the Longroom, the port's nerve centre located in a tower just west of the entrance to Millbay Docks, VHF call sign *Longroom Port Control*, channels 16 and 14.

Traffic light signals are displayed to seawards from the conspicuous tower on Drake's Island and inland from Flagstaff, Devonport, on the eastern shore of the River Tamar just upstream of Torpoint ferry. These signals control shipping movements in Drake Channel, the Narrows and the Hamoaze, the wide stretch of water running inland west of the city where the Naval Dockyard is situated:

a) Unlit = no restrictions to movement.
b) Three vertically flashing red lights = emergency, all traffic movements suspended unless directed otherwise by Port control.
c) Occulting red light vertically over two occulting green lights = only outgoing traffic may proceed in recommended

channels; crossing traffic to seek permission from Port Control.
d) Two occulting green lights vertically over occulting red light = only only ingoing traffic may proceed within recommended channel; crossing traffic to seek permission from Port Control
e) Two occulting green lights over occulting white light = vessels may proceed in either direction but to give wide berth to H.M. vessels.

If there are no traffic signals in force, from sunrise to sunset, a wind strength warning light will be diplayed if necessary, (Occ) = wind force 5-7. (2 Occ vert) = wind greater than force 7.

Anywhere in the buoyed channels north of Plymouth breakwater Colregs Rule 9 applies: *'....in narrow channels keep to starboard whether under power or sail. A yacht under 20m in length must not impede larger vessels confined to a channel.'*

In practice, there is no reason why small craft should get embroiled with big ship movements in Plymouth as for the most part there is no need to follow any of the buoyed channels. Keep well clear, and let common sense and good seamanship prevail.

Hazard number two is the *HMS Cambridge* gunnery range at Wembury which fires to seaward, usually from 0930-1600 Tuesdays to Fridays (occasionally longer in summer); when active four large red flags are flown at Wembury Point. Firings involve anything from 7.62mm to 4½inch shells, and are very audible. The outer limits of the range cover a sector of 130°T to 245°T from Wembury Point extending approximately 13 miles to seawards. This will be extended to 17 miles in August 1995 in preparation for the transfer of Naval Sea training to the area when Portland Naval Base closes in 1996, which will make the range and approaches to Plymouth much busier.

The main surface targets, usually large orange floats towed by *AV Seawork* launches, tend to be 6-7 miles offshore; air targets, towed behind light aircraft, approach from 7 miles to within 1 mile of the land.

Pleasure craft pose a major problem as they can often delay firing and it is greatly appreciated if you can, ideally, avoid the range altogether, or, if on passage, keep within 3M of the shore and clear the area as fast as possible.

The range monitors VHF channel 16, and works on 11,12 and 14, and it helps them considerably if you can call

Start Point to Rame Head

Wembury Range to identify yourself and advise of your destination. If there seems to be nothing happening although the red flags are up, it means they are hoping to fire and it could just be you that's stopping them. If you wish to phone ahead for firing details, ring Plymouth 862779 or 553740, extension Cambridge 77412, or Freephone 0800 833608

The large open roadstead of Plymouth Sound was transformed as a fleet anchorage by the completion of Sir John Rennie's central breakwater in 1841. Just under a mile long, it consumed 4½ million tons of materials, and took 29 years to build. Entry is easy round either end; from the east, after passing the Great Mewstone, the Shagstone, a rock marked with a white beacon, is the only offlying hazard on the eastern shore, and it can be passed reasonably close, keeping you well east of the deep water channel, which is bounded on its western side by the Tinker shoal, least depth 3.5m LAT.

This is marked on its eastern extremity by the 'East Tinker' BYB east cardinal buoy (Q (3) 10s) to the west by the 'West Tinker' YBY west cardinal buoy (VQ(a)10s) Although normally not a problem, seas break heavily on this shoal in southerly weather, when the western entrance is far better. Whidbey Light (Occ (2) WRG 10s 5M) on the eastern shore is the first of several complex sectored approach lights for the eastern deep water channel, the next, Staddon Point, east of the breakwater end, has a light beacon (Occ WRG 10s 8M) with R/W horizontal bands around its daymark base.

The eastern breakwater end is marked by a conical daymark surmounted by a beacon with round ball topmark and a sectored light (Iso WR 5s 9m 8M) showing: Red 001°-018°, White 018°-190°, Red 190°-353°, White 353°-001°

The East Tinker buoy lies in the narrow white sector and you soon pass into the red. Large ships will then come onto the leading lights situated on and to seaward of the Hoe, the upper (Occ G 1.3s), the lower sectored (Q WRG), the white sector covering the leading line of 349°T.

Once past the breakwater end a YBY beacon with west cardinal topmark (Q (9) 15s) marks the eastern side of the deep water channel; closer inshore (2FG vert) lights mark the outer end of Fort Bovisand pier. There are several large Admiralty moorings inside the breakwater, all of which show (Fl Y) lights with intervals of between (15s) and (2s).

Approaching from the west, once past Rame Head keep to seaward of the 'Draystone' red can south-east of Penlee Point (Fl(2)R 5s) and then steer straight for the lighthouse on the western end of the breakwater (Fl WR 10s, Iso 4s over sector 031°-039°). There are several Admiralty buoys in this approach, of no navigational significance, which display (Fl Y) lights. However, in daytime, from well offshore the breakwater ends are not immediately easy to spot; the large round fort in the centre is far more conspicuous.

Cawsand Bay lies on your port hand providing an excellent anchorage in south-west and westerly weather, off the pleasant twin villages of Cawsand and Kingsand. This is a handy overnight stop on passage along the coast to avoid the detour into Plymouth and has the additional

The anchorage in Cawsand Bay is well sheltered from the west and always popular at weekends

advantage of being free; anchor anywhere clear of the local moorings where there is about 2.5m LAT with good holding. At the weekend this is a very popular spot for local boats and often crowded.

Ashore the facilities are surprisingly good. No less than five pubs, all providing meals, three restaurants and a bistro. There are also three general stores/off-licences, two of which are open until 2100 in summer, and the owners of the 'Shop in the Square' in Cawsand will allow you to fill water containers from their tap. 'Shipshape' in Kingsand has a limited amount of chandlery, Calor and Gaz refills and it also houses the local Post office. There is also a butcher (excellent home-made sausages) a baker and a public phone up the road to the left of the prominent Kingsand clock tower.

There are summer ferries to and from the Barbican, which is within easy walking distance of Plymouth City centre. If you follow the delightful coastal footpath through the Mount Edgcumbe Country park, Cremyll ferry to Plymouth is just over 3½ miles away – about an hour's walk. It is advisable to leave someone on board if such an expedition is planned in case the wind hooks round to the east when you should get out fast and into the shelter of Plymouth before Cawsand becomes untenable.

Approaching Plymouth from offshore, the most distinctive features are the large grain silo by Millbay Docks (rather like a huge white cathedral) and Drake's Island to the west; the large square Hotel building in the centre, and to the east the high ground of Staddon Heights, topped with radio masts. There are two prominent forts on either side of the Sound just inside the breakwater, Picklecombe on the western shore, now luxury flats, and Bovisand, opposite, the British Sub-Aqua Club diving training centre. Two of 'Palmerston's Follies', they were built as part of a defensive chain along the south coast in the 19th century in anticipation of a French invasion that never materialised.

The thing to decide early is where you intend to berth, as the **Mayflower International Marina** is to the west of the town centre; **Sutton Harbour Marina, Queen Anne's Battery Marina** and **Clovelly Bay Marina** are to the east. The other alternatives depend very much on the prevailing weather conditions – either anchoring, or picking up a mooring, although here the choice is very limited.

The **anchorages** in the Sound are all free, but suffer from the disadvantage that they are remote from the centre of Plymouth. The southern end of **Jennycliff Bay**, protected by Staddon Heights, provides good shelter in easterly weather, sounding in to about 2m at LW well clear of the conical green buoy (Fl G 6s) marking the wreck of the *MV Fylrix* which dries LAT. The inshore area to the north of the buoy, across Batten Bay, is a water skiing area.

Both **Drake's Island** and **Barn Pool** provide shelter in west or south-westerly conditions and can be approached either by following the main deepwater channel round to the east of Drake's Island, or west of it by using the short cut across the **'Bridge'** providing you have sufficient rise of tide.

Shallows lie on both sides of this narrow passage, and on them the dangerous remains of wartime concrete obstructions and dolphins. However, between them there

Barn Pool. Note the Great Mewstone in the distance

is 1.7m LAT, and the channel over the Bridge is clearly marked by four lighted beacons, the outermost No 1 green, with a green conical topmark (Q G), which should be left on your starboard hand, No 2 a red port hand beacon with red can topmark (Q R), No 3 green with green conical topmark (Fl (3) G 10s) and the innermost, No 4 red with red can topmark (Fl (4) R 10s). No 1 and No 4 have tide gauges, graduated in metres, indicating the height of tide above CD.

At Springs, the stream can run through here at 3 knots; use the passage with caution, ideally no earlier than an hour or so after LW, do not wander from the channel, and unless you have a good following wind, proceed under power.

Once through, Barn Pool lies to port, under the wooded Mount Edgcumbe shore. Sound in to about 3-4m LAT, the deep water carrying close to the pebbly beach. A trip line is recommended as there are many underwater obstructions and the southern end of the bay is foul, supposedly with the remains of an old sunken barge. Although eddies extend into the pool at times you will lie here comfortably out of the main tidal stream.

This is a delightfully peaceful spot surrounded by the magnificent woods and grasslands of the 800 acre Mount Edgcumbe Country Park which is open to the public daily, along with Mount Edgcumbe House, a restored Tudor mansion. There is also a fine coastal footpath through the park which eventually takes you to Cawsand.

You can either stroll through the park, or, two hours after HW, along the foreshore to Cremyll, for a pint at the Edgcumbe Arms which also does good food. Close by you will find the famous Mashford Boatyard where Chichester and Rose both fitted out for their circumnavigations. An old family business, this traditional yard has a long held reputation for wood building and repairs. A regular passenger ferry runs from Cremyll to Stonehouse from where there are buses right into the centre of Plymouth.

The only other feasible anchorage lies north-east of Drake's Island, east of the pier, well clear of the local moorings and the underwater obstruction, least depth 0.9m LAT 400m due north of the pier. Reasonable shelter will be found in winds between south and west but the proximity to the busy main channel can make it rolly at times.

Once known as St Nicholas's Island, this is traditionally

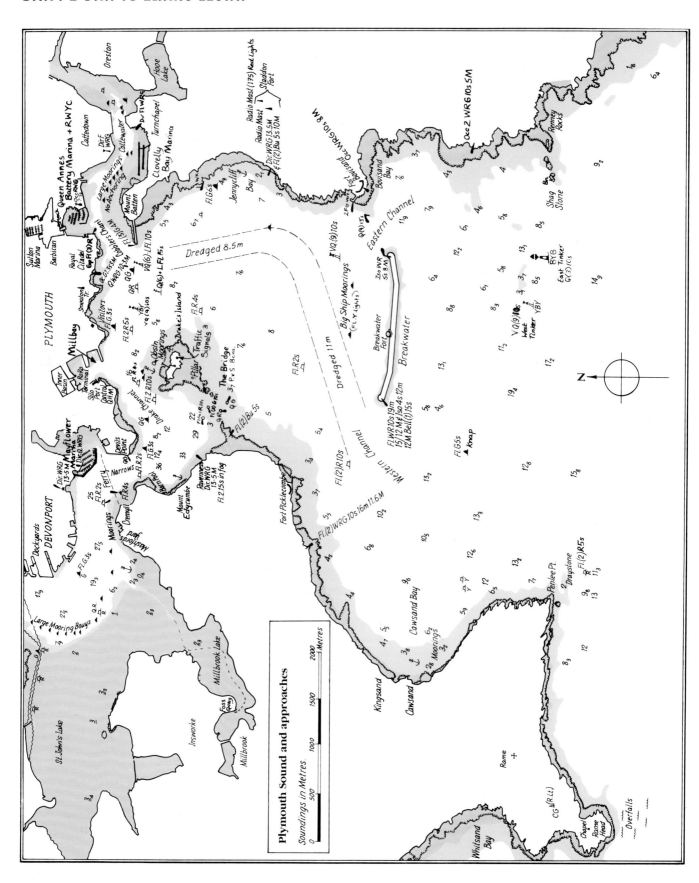

where Francis Drake lay with his battered *Golden Hinde* on the return from his circumnavigation in 1580, cautiously waiting for news of the political climate that might greet him, before sailing on to be knighted at Deptford by Queen Elizabeth 1.

Drake's Island is Crown Property and has been used for a variety of purposes including adventure training but is now uninhabited apart from a caretaker and landing is not advised due to the dilapidated pier and buildings.

The Mayflower International Marina

With few moorings or anchorages and a choice of four

Mayflower International Marina has easy access. Devil's Point is on the left, Cremyll opposite, and the narrows broaden upstream at the mouth of the Hamoaze

marinas most visitors to Plymouth tend to opt for the convenience – and expense – of one of them. The Mayflower International Marina, the only one in the south-west to be awarded Five Gold Anchors by The Yacht Harbours Association, lies to the west of Plymouth, and is approached either by the Bridge if conditions permit, or along the deep and well-lit Drake's Channel leaving the conical green 'W Vanguard' (Fl G 3s) buoy to starboard and through the narrows between Cremyll and the orange and white beacon on Devils Point (QG). From here the modern buildings of Ocean Court and the forest of masts in the marina are unmistakeable – either call *Mayflower Marina* on VHF channel 80 or 37 for a berth allocation and proceed straight to it or tie up in the reception berth on the outer pontoon which is clearly marked between the striped flags and report to the office. At night the marina has an approach light (Dir Q WRG) and the south-eastern end of the outer pontoons is marked by (2 FR (vert)).

Much of the Mayflower's success has undoubtedly been due to the fact that it is owned by a consortium of its berthholders. It has a particularly relaxed, quiet and friendly atmosphere, and as a popular stop for long distance cruisers, a genuinely International flavour. Although its position away from the main centre of Plymouth would seem to be a disadvantage it is in fact quite self-contained and the facilities are excellent, with a grocery, chandlery, restaurant,

off-licence, luxury showers, toilet, launderette, and a clubhouse/bar which welcomes visitors. During the season a courtesy bus runs daily into the centre of Plymouth and other services include diesel, petrol and gas (available 24 hours) and electricity; weatherfax, marine and electronic engineers, rigging, divers, a crane and 25 ton travelift, and the excellent 24 hour security makes it an ideal place to leave a boat for longer periods of time with special rates available.

There are 270 deep water berths taking up to 34m LOA and 4m draught, with about 30 for visitors although even at the height of the season they always manage to fit you in somehow. The overnight charge for a 30 footer is £14.85 a night inc VAT, with no surcharge for multihulls.

Approaches to Queen Anne's Battery Marina, Sutton Harbour and Marina, the Cattewater and Clovelly Bay Marina

Approaching from seawards and passing to the east of Drake's Island, **Plymouth Hoe** runs east-west, a high grassy slope with the old Eddystone lighthouse, banded horizontally red and white, and the tall obelisk of the war memorial both prominent. At its eastern end the large fortress of the Royal Citadel and the Royal Plymouth Corinthian Yacht Club beneath it overlook the western side of the entrance to the Cattewater and Sutton Harbour. The

RPCYC welcomes visitors and has two daytime moorings off the clubhouse and two visitors' moorings in the Cattewater available at £5 a night. The clubhouse is open daily 0900-1500, 1800-2300 (closed all day Monday and Sunday evenings) and monitors VHF channel 37; showers, bar and snacks are all available.

Directly opposite, Mount Batten, an isolated hill and former RAF air/sea rescue base, forms the eastern side of the entrance with a breakwater extending towards an outwardly confusing cluster of buoys marking the deep water channel and a large dolphin on the Mallard Shoal, least depth 3.5m LAT. This has a triangular white topmark surmounted by a sectored light (Q RWG) and is the lowest of the two leading marks for the main deep water channel, the upper, a beacon with white triangular topmark and (Occ G) light is on the eastern side of the Hoe giving a transit of 349°T.

In practice there is no need for small craft to follow this line; pass between the 'South Mallard' YB south cardinal buoy (VQ (6) + Fl 10s) and the end of Mount Batten breakwater (2FG (vert)) and into the Cobbler Channel. Ahead, **Queen Anne's Battery Marina** (known locally as 'QAB') breakwater is fronted by vertical piling and has a orange/white horizontally striped daymark with orange spherical topmark on its southern side. Both QAB and **Sutton Harbour** are entered past Fisher's Nose, a granite quay on the western shore with 'Speed Limit 8 knots to the east' painted on it and a light (Fl (3) R). You must proceed under power beyond Fishers Nose as it is always a busy corner; most of the large trip boats operate from here, and it is the entrance to two marinas and the fishing harbour.

The River Plym continues eastwards, and its mouth, known as the **Cattewater**, is commercial and busy with coaster traffic and administered separately by the Cattewater Harbour Commissioners. Due to the local congestion anchoring is prohibited within the Cattewater. Plymouth's newest marina at **Clovelly Bay** is situated on the southern shore of the Cattewater at Turnchapel, just under ½ mile upstream. A further ½ mile on, the road bridge at Laira has 5m clearance which effectively closes the river to all but small craft.

Queen Anne's Battery Marina

Keep to the starboard side of the channel and follow the marina breakwater to enter, watching out for and giving way to boats emerging as they have priority. At night there is a sectored approach light (Occ 7.5s RWG) which is situated in the small clock tower on top of the marina building/RWYC clubhouse – the white sector giving the approach through the Cobbler channel. The breakwater has an (Occ G) light on the south-western corner, and a (QG) light on the breakwater end. Queen Anne's Battery Marina can be contacted on VHF channels 80 or 37 to reserve a berth, call sign *QAB*. Sixty are normally available for visitors alongside the continuous pontoon on the inside of the breakwater and at the height of the season you will probably have to raft up. If you have no VHF, berth anywhere here and check in at the marina office. The finger piers are all reserved for permanent berth holders but may be allocated to visitors if berth holders are away.

QAB and Sutton Harbour with access lock, centre left

The marina opened in 1986 and took its name from the former use of the site as a gun emplacement built during the Napoleonic War. Today there are comprehensive onshore facilities, including a chandler, an excellent new and secondhand maritime bookshop and Admiralty Chart Agent and many other marine orientated businesses from liferaft repairs to stainless steel fabrication, diving air to sailmakers.

There is water and electricity on the pontoons, diesel and petrol are available from the fuel barge (0830-1830) on the outer pontoon adjacent to the marina entrance. All the pontoons have security gates; ashore there are showers and toilets, a launderette, cafe and Marina clubhouse/bar. There is also a large slipway and 20 ton travel hoist for repairs. The overnight charge for a 30 footer works out at £15.86 inc VAT, with no surcharge for multihulls.

In early 1989 the prestigious Royal Western Yacht Club of England moved to its smart premises within the marina after 25 years at the west end of the Hoe, in the building that is now the Sir Francis Drake Pub. This and the enthusiasm of the marina's Managing Director Mark Gatehouse – a well-known offshore racer – has established QAB as a major venue for international yachting events, most of which are organised by the RWYC, such as the Round Britain and Ireland Race, Fastnet finish, Twostar and, of course, the Singlehanded Transatlantic.

Visitors berth on the long inner pontoon at QAB

The fifth oldest club in England, founded in 1827, the RWYC has four visitors' moorings in the Cattewater available on application at around £7.50 a night and visitors are welcome to the use the Club which is open 0900 – 2300 with showers, bar and restaurant.

During the big race events space at QAB is inevitably at a premium and it is unlikely that you will find a berth without a prior booking. Another slight drawback is the marina's position opposite the main nautical centre of Plymouth, the Barbican, and the rest of the city. The surrounding hinterland is rather run down and industrial, but the new lock and bridge across the entrance to Sutton Harbour now means that it is now a much shorter walk to the Barbican than previously – turn left as you leave the main marina entrance and follow the road and footpath.

It is possible to dinghy over and land by the Mayflower Sailing Club which also welcomes visitors with a bar and showers available, but by far the simplest solution is the marina water taxi at £1 a head which lands and picks up near the Mayflower Steps on request.

Sutton Harbour and Marina

Sutton Harbour is the home of the Plymouth fishing fleet, a large number of sea-angling boats and Sutton Marina, the first to be built in Plymouth, in 1973. Privately owned by the Sutton Harbour Company, the harbour is entered through a 44m long x 12m wide lock opened in 1993 as part of the Barbican flood prevention scheme. This main-

Sutton Harbour is entered through a lock

tains an approximate depth of 3.5m above CD within Sutton Harbour and entry/exit is available free of charge 24 hours daily. Towards HW the gates remain open to allow a freeflow although the footbridge still has to be opened to all but small craft with no masts. The lock and footbridge are opened on request by the Port Control Office which overlooks the lock on the eastern side, call *Sutton Harbour Radio* direct on VHF channel 12 or, lacking VHF, stand by in the immediate vicinity, noting the traffic light signals displayed from the lock entrance: *3 vert red = STOP. 3 vert green = GO. 3 flashing red = serious hazard, WAIT.*

Transit is made easy with floating pontoons for temporary berthing inside the lock. At night, the lock is floodlit and the entrance is indicated by reflective chevrons, R/W to port, G/W to starboard.

Once inside Sutton Harbour the old fishmarket and Barbican quays lie on your port hand, the new fishmarket is to starboard and Sutton Marina is dead ahead, with the visitors' arrival berth clearly marked on the outer end, adjacent to the fuel pontoon. At night the eastern end of the outer pontoon displays a (QR) light.

There is a total of 310 berths and subject to availability they can usually accommodate between 25 and 30 visitors of average size and up to a maximum of 19m LOA if you contact them in advance. As elsewhere, it is always best to call ahead during the season.

The Marina Office is located right at the end of Sutton Pier, overlooking the fuel pontoon, and a 24 hour VHF watch is maintained on Channels 16, 80, 37 and 12, call sign *Sutton Harbour Radio.*

On-site facilities include showers, toilets, a small launderette, and 24 hour security. A number of marine businesses, including electronic engineers and chandlery, are located on Sutton Pier beside the marina; shops and all the facilities of the Barbican are a short walk around the quayside. Water is available on the pontoons, and the fuel berth, petrol and diesel, is open daily 0830-1830 in summer; daylight hours in winter. Charges for a 30 footer are £15 per night, inclusive of VAT, with reduced rates for longer stays. Close by on the eastern side of the Harbour the comprehensive boatyard of Harbour Marine has a 25 ton crane, slipway and full repair facilities.

The Barbican

The tourist centre of Plymouth, the bustling and historic area of the Barbican, is reputedly where the Pilgrim Fathers embarked aboard the *Mayflower* and sailed for the New World in 1620. The Naval city of Plymouth was blitzed more heavily than anywhere else in Britain during the last War, and large areas were completely destroyed. The Barbican alone remains as an outstanding example of what this medieval city was like before 1941, a maze of intricate narrow streets, and fine examples of Tudor buildings, many of them now shops, restaurants and pubs in this interesting and attractive area surrounding Sutton Pool. Several are museums, including the Elizabethan House, in New Street, and nearby, the 16th century Merchant's House in St Andrew's Street which houses the Museum of Plymouth History, open weekdays 1000-1300, 1415-1730, 1700 on Saturdays.

Clovelly Bay Marina

The 180 berth Clovelly Bay Marina which opened off Turnchapel in 1990 usually has between 18 and 20 visitor's berths up to 45m LOA available on the long outer pontoon, which has easy all-tide access from the Cattewater and over 6m alongside even at MLWS making it particularly attractive for owners of larger vessels. Call *Clovelly Bay* on VHF channel 37 or phone in advance; prices for a 30 footer work out at £12 per night inc VAT; with special rates for longer stays, it can be a useful place if you have to leave a boat for a while.

Facilities include diesel (alongside HW only, otherwise in cans) water, electricity, Calor/Gaz, toilets/showers, 24 hour security, on-site launderette, and repairs including

Clovelly Bay Marina. Entrance to the Cattewater on the left

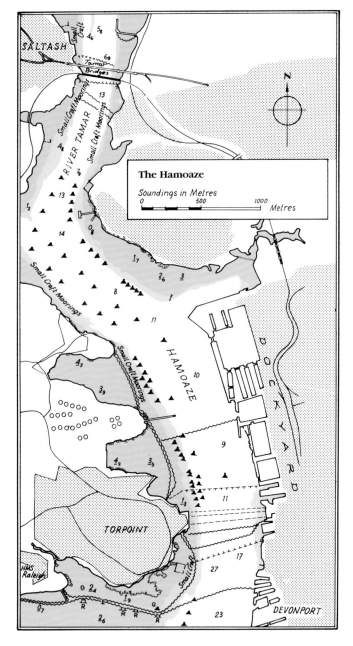

marine engineer. There is also a basic chandlery/food store, public phone just outside the marina entrance; close by, the 'Boringdon Arms' and 'New Inn' both serve meals.

Although seemingly less conveniently situated for a visit to Plymouth there is a regular ferry service to the Barbican and a water taxi is also available on request.

The City of Plymouth

Plymouth has a population of over 250,000, and the city centre, ten minutes' easy walk from the Barbican, can obviously provide anything that might normally be required. The post-war redevelopment of the main shopping area surrounding Royal Parade is mostly redbrick, and architecturally unimaginative, with wide boulevards, and rather demanding shopping precincts; somehow one feels a great opportunity was lost in the rebuilding. The main line railway station is to the north of the main centre, and the Bretonside bus and coach station lies between the Barbican and the city centre.

However, in contrast to the mundanity of mid-20th century planners, the seafront is a grand spectacle. It is a short walk from the Barbican, following Madeira Road up past the Citadel, and the elevated promenade winds beneath the grassy slopes of the Hoe, overlooking the rocky foreshore and cliffs below, which are dotted with bathing platforms, pathways, and the large open air swimming pool.

Climbing higher onto the open space of the Hoe, the views across the Sound from this natural grandstand are magnificent, and even more so if you pay the extra 25p to climb to the top of Smeaton's former Eddystone lighthouse, built in 1759, dismantled in 1882, and rebuilt on the Hoe in 1884.

The Hoe, bowls and Sir Francis Drake will forever remain synonymous, a seemingly remarkable display of sang-froid, contentedly playing on as the vast Spanish Armada sailed unchallenged into the Channel. Far more a seaman than a bowls player, Drake, of course, knew only too well that his ungainly ships could not leave the Sound against the head wind until the ebb began . . .

The Hamoaze and beyond

One of the real advantages of Plymouth if the weather turns against you is the great potential for exploring further

inland along the Rivers Tamar and Lynher. Heading west past the Hoe, **Millbay Docks** is commercial and of no interest to visitors. It is the terminal for the RoRo ferries to Roscoff and Santander, and care should be taken to avoid these large vessels entering and leaving. On the starboard side of the docks entrance the small cluster of masts belongs to boats moored in **Millbay Marina Village** which is private and has no facilities for visitors. Plymouth lifeboat is based within the marina.

Beyond the Narrows and the Mayflower International Marina you enter the wide **Hamoaze**, its curious name derived from the thick mud that once oozed out of Ham Creek, long since buried beneath the extensive Royal Naval Dockyards that line the eastern, Devonport, shore. Established by King William III in 1691 these are now privately run by Devonport Management Ltd (DML), who are better known in the sailing world for the fleet of 10 identical 67 foot racing yachts that were built here between 1990 and 1992 for Chay Blyth's around-the-world British Steel Challenge.

This is always a fascinating stretch of water, particularly for the younger crew members as there are invariably a variety of warships and submarines berthed along the quays. For the Skipper, however, it is a more demanding exercise as it is usually busy and a careful eye should be kept on other ship movements; remember, too, that civilian craft are not permitted to pass within 50 metres of military vessels, Crown Property or enter the Dockyard basins.

The large figurehead of the founder, King William III, at the southern end of the dockyards is known locally as 'King Billy'. Behind him, the large covered slipway is the oldest in any of the former Royal dockyards, and as you pass upstream, the first group of three huge sheds is the undercover frigate repair facility, the next massive building with a very large crane is the complex where Britain's nuclear submarines are refitted.

Large chain ferries link Devonport with Torpoint on the Cornish shore, and until 1962 when the Tamar road bridge was opened at Saltash, a mile further upstream, this was the only road link across the river. The ferries have right of way and display flashing orange lights at their forward end to indicate the direction in which they are moving; if in doubt always pass astern and give them a good berth.

In contrast the western shore is much less developed, and beyond Mashford's boatyard the shallow inlet of Millbrook Lake stretches away to the west for over a mile to the village of Millbrook, making an interesting diversion for shallower draught boats on the flood, ideally 3 hours after LW.

Initally the channel heads directly for the buildings and slipways at Southdown Quay on the northern side of the creek where there are usually a number of fishing boats and pleasure boats on pontoons and moorings. Leave these on your starboard hand and then steer south-east towards the prominent house beneath the woods on the southern shore which has the road running in front of it. As you close this bear west again, leaving both the black post with yellow X topmark and the line of boats on green mooring buoys on your starboard hand, at which point you will see a long pontoon on your starboard bow extending from the Multihull Centre boatyard at Foss Quay. If there is space on the pontoon berth here and make contact with the office, otherwise berth alongside the quay. In both places you will dry at LW.

Visitors are very welcome to stay for up to 2 nights free of charge, after which you will have to pay a weekly rate of about £17. The yard can provide a shower and toilet, chandlery, repairs, diesel in cans, water and rigging, as well as a crane. Millbrook is within easy walking distance with a good selection of general stores, a Post Office and three pubs which all do food.

Torpoint Yacht Harbour is located on the western side of the Hamoaze just south of the ferries, in the old Ballast Pond. This walled compound is accessible 3 hours either side of HW and the berths are all taken up with local boats.

The Lynher/St German's River

The Lynher, or St German's River is the first opportunity to get away from the bustle of the Hamoaze, and its wide mouth opens to port, upstream of the dockyards, beyond the large warship moorings. The river dries extensively, but is navigable on the tide for four miles to the private quay

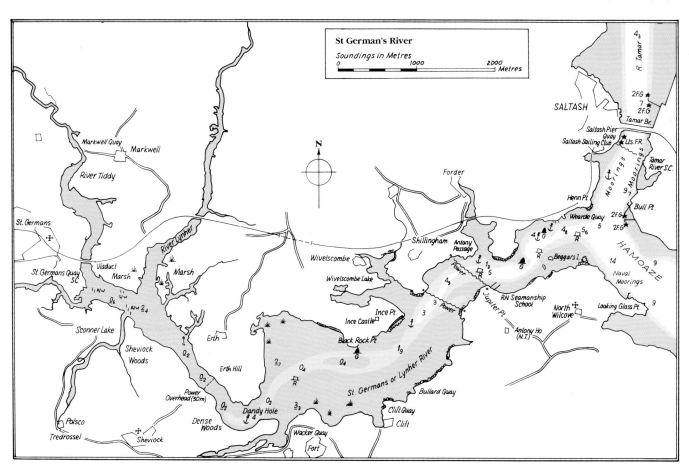

Start Point to Rame Head

Dandy Hole – 'a fortuitous pool'

at **St German's** where it is possible for bilge keelers to dry out or larger boats to lie alongside by arrangement.

There are a few buoys in the lower part of the river; red cans should all be left to port and conical green buoys to starboard. The channel enters along the northern side, past Wearde Quay, where there are local moorings, and the first two port hand buoys clear Beggars Island, a gravelly shoal awash at HWS on the southern side of the entrance. Just east of Sand Acre Point, and the first green buoy, there is a possible anchorage, but the proximity of the main line railway tends to disturb the potential peace. The channel trends south to avoid the spit extending from the northern shore, marked by another green buoy, passing a number of moorings belonging to the R.N. School of Seamanship at Jupiter Point, a wooded promontory with a jetty and pontoons. The next red buoy is close to the northern shore, off Anthony Passage, where there is a pub, and it is possible to anchor off the mouth of Forder Lake. West of Forder Lake as far as the next creek, Wivelscombe Lake, an underwater power cable and gas pipeline cross the river and anchoring is not advisable.

Ince Point, on the western side of Wivelscombe Lake is surmounted by Ince Castle and south of it there is another anchorage in about 3m LAT. At high tide this is a broad expanse of water, surrounded by lush fields and gently rolling hills, peaceful and unspoiled, with very little evidence of human intrusion. However, from here on depths reduce considerably, with generally less than 1m LAT and drying banks are extensive on both shores. The various salty creeks are ideal for dinghy exploration, but on the flood, with an eye on the sounder, the river should be explored further, for the best is yet to come. Keep close to the next green buoy off Black Rock Point, where the channel narrows considerably, and at the next red can the shores ahead close, becoming steep and wooded along Warren Point, and the river disappearing as it turns tightly behind the rounded slope of Erth Hill. Here, between Warren Point and the bend there is a fine and isolated anchorage in **Dandy Hole**, a fortuitous pool with about 3m LAT. Overlooked by high woods, this is as peaceful a spot as you are likely to find, well sheltered and totally away from it all.

From Dandy Hole to St German's the river effectively dries completely and is easiest to explore by dinghy, the wooded shores opening out again and the twisting channel is marked by red posts to port, green to starboard, swinging from the eastern shore above Erth Hill to the western off Sconner Lake, then north to a striped middle ground pole where the Rivers Lynher and Tiddy part company, the latter heading west towards St German's Quay which is dwarfed beneath the railway viaduct across the river. The channel of the Lynher continues north, marked by occasional white posts, before passing under another railway viaduct (21m clearance) beyond which 'Boating World' is situated with a large number of second hand craft, sailmaker, chandler, marine engineer, 10 ton hoist and 35 ton crane. (Tel: 851679). Normally this is accessible for boats of moderate draught from half flood onwards.

At St German's there are a number of drying moorings off this private quay, and other local boats berth alongside

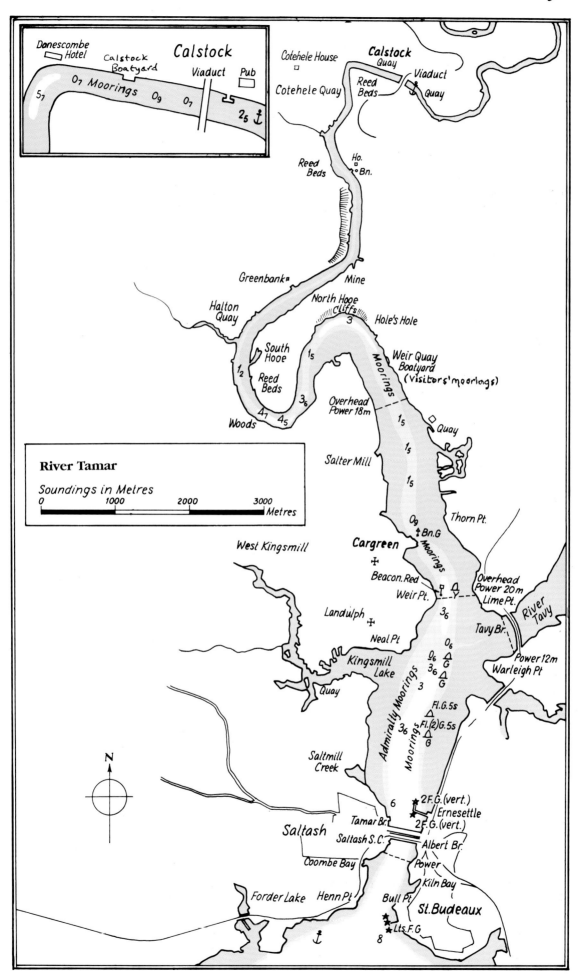

Calstock

Danescombe Hotel
Calstock Boatyard
Viaduct
Pub
O₇ Moorings O₉ O₇
5₇
2₅ ⚓

Cotehele House
Calstock Quay
Cotehele Quay
Reed Beds
Viaduct
Quay

Reed Beds

Ho.
Bn.

Greenbank
Mine

Halton Quay
North Hooe Cliffs
Hole's Hole
3
South Hooe
Weir Quay Boatyard (visitors' moorings)
Moorings
1₅
Reed Beds
1₂
3₆
Quay
Woods
4₇ 4₅

Overhead Power 18m
1₅

Salter Mill
1₅

1₅

River Tamar

Soundings in Metres

0   1000   2000   3000
Metres

O₉
⚓ Bn. G

West Kingsmill

Thorn Pt.

Cargreen
Moorings
Overhead Power 20m
Lime Pt.
River Tavy

Beacon. Red
Weir Pt.
3₆

Landulph
Tavy Br.

Neal Pt.
O₆
Power 12m
Warleigh Pt.

Kingsmill Lake
O₆
3₆ G
3

Quay

Admiralty Moorings
Fl.G.5s
G

Saltmill Creek
3₆
Fl.(2)G.5s
G

Moorings

N

6
2 F.G.(vert.)
Ernesettle

Saltash
Tamar Br.
2 F.G.(vert.)
Saltash S.C.
Albert Br.
Coombe Bay
Power

Kiln Bay
Forder Lake
Henn Pt.
Bull Pt.
St. Budeaux
⚓
Lts. F.G
8

Start Point to Rame Head

St German's Quay is private but it is sometimes possible to dry out alongside by arrangement with the Sailing Club

on the soft mud. Downstream of the old warehouse buildings where the Quay Sailing Club is based no berthing is allowed, but upstream, space can sometimes be found for an overnight stay alongside by arrangement with the club. There is a noticeboard with a phone number to contact when the club is closed; normally it is open on Wednesday evenings and at weekends. There is a small dinghy landing pontoon alongside the quay and a water tap by the club.

This is a peaceful corner, just a few cottages and the old grassy quayside disturbed only by the rumble of the trains overhead. The village is about ½ away where there are basic provisions, pub, and Post Office; you will pass a phone box en route. Occasional trains run to and from Plymouth.

The River Tamar

The Tamar, however, is altogether a far bigger proposition for it is twelve miles to **Calstock**, and ideally a full flood tide is needed. In the narrow upper reaches the streams run strongly particularly on the ebb, in excess of 5 knots when the river is in spate. Due to the current, and the amount of floating debris including trees and branches, in these circumstances such a trip is not recommended.

Today, the river is a tranquil place, a quiet rural waterway that betrays little indication of its former importance as one of the busiest industrial areas in the West Country. Extensive granite quarries, and rich tin, copper, silver and arsenic mines in the upper reaches were all serviced from the sea, with large sailing schooners, ketches and barges plying far inland. Apart from a few overgrown and derelict quays, it is now difficult to imagine such a hive of maritime activity; fortunately, the heritage has not been totally lost; at Cothele, far upstream, there is a restored Tamar barge and a museum, and Morwellham, almost at the head of navigation, is a former Victorian port that has been restored as a tourist attraction.

Above the Lynher, the channel is broad and deep past **Saltash** on the western shore, beneath the twin road and rail bridges. The latter, yet another of Isambard Kingdom Brunel's remarkable achievements, took seven years to build and was opened in 1859 to carry the Great Western Railway from London into Cornwall.

There are many local trot moorings off Saltash, and a slipway where it is possible to land just upstream of the Quay where the Saltash Sailing Club is based; apart from this there is little else here of interest to visitors.

Continuing upstream beyond the bridges on the eastern shore is Ernesettle Pier, an MOD munitions depot, and the two long lines of Admiralty trot moorings running upstream with barges upon them lie on either side of the main channel.

A line of four green conical buoys marks the eastern bank, and the northernmost lies off the entrance to the River Tavy. Sadly this attractive tributary is closed to sailing boats by the railway bridge across its mouth with 7.6m clearance; the river dries almost completely at LW, but power boats or dinghies can explore it on the tide, as far as Bere Ferrers, where there is a pub and limited provisions.

Between the buoy and Neal Point, on the western shore, there is a drying bank, keep close to the starboard side of the channel, before steering north to Weir Point, a wooded promontory where a large overhead power line crosses the river. Almost directly beneath it, the channel is marked by a a post with square red topmark to port, and a yellow conical buoy to starboard. Ahead lies the small village of Cargreen, with a number of local moorings which indicate the deeper water.

Cargreen

Before the first world war Cargreen was still a major crossing point for the Tamar but the growth of motor traffic resulted in the enlargement of the Torpoint ferries, and the decline of Cargreen. Today it is another peaceful village, not much more than a single street running down to the quay. There is a Post Office, closed on Tuesday afternoons, and a general store which closes on Friday afternoons. Anchor clear of the local moorings where there is about 2m LAT, and no charges. Cargreen Yacht Club has no facilities for visitors except a tap but you can land at the club slip which is particularly convenient at low tide. They will probably be able to find you a mooring, too, if members are away.

Beyond the moorings the channel is marked on the starboard side by a green beacon with triangular topmark, and it is a straight run northwards to Weir Quay with moorings lying along both sides of the deeper water. Here, there is a boatyard with a 12 ton crane and full repair facilities with visitors' moorings (£5 a night), diesel in cans and showers.

Continuing upstream, apart from a few deeper pools, there is less than 1.5m LAT, the river narrows considerably, and winds almost back on itself. The channel is not marked, but generally the deepest water lies along the outside of the bends.

Keep close to the wooded shore beyond Holes Hole, where there is an old quay and several hulks, following the low cliffs right round the outside of this bend as the bank extending from the south shore is very shallow, the few moorings lying along its edge. From here onwards, between March and September, salmon netsmen will be encountered and care should be taken to pass slowly,

keeping a lookout for the nets extending from the banks. As the extensive reed beds come abeam to starboard head across to the southern shore, and again hold close to the steep woods.

Large trip boats run regular day cruises from Plymouth to Calstock depending on the tides and if encountered in these upper reaches there is not a great deal of water to spare. Their Skippers appreciate it greatly if you can pull over to let them pass.

The channel continues to follow the western bank with Pentillie Castle looking down from the hillside and past Halton Quay with a curious building rather like a railway signal box that is, in fact, one of the smallest chapels in England. Bearing across the the eastern shore past extensive reedbeds, upstream, there is a prominent house with two gables, and on the bank a post with triangular white topmark which should be kept in line with the left hand gable to hold the channel. Beyond the post, the deeper water swings back to the western bank, then midstream as Cothele Quay appears.

Cothele Quay

With sufficient water it is possible to berth temporarily alongside the quay which has been preserved by the National Trust as part of the Cothele estate, and is home for the restored Tamar barge *Shamrock*. A 57 foot ketch, she was built in 1899 and is co-owned by the National Maritime Museum. The only surviving example of the barges that were once an essential element in the life of the waterway, during the summer she is occasionally sailed by enthusiasts. There is a small museum on the quay devoted to the maritime history of the river, and about ten minutes' walk from the quay Cothele House, a splendid Tudor mansion and gardens, are open daily April – October, 1100-1800.

Calstock

Past the moorings off the Quay, the prominent building of the Danescombe Hotel with its elegant veranda stands on the hillside above the final sharp bend into the Calstock reach, and as the reed bends open to starboard, beyond them the viaduct will come into view. A line of moorings lies in the centre of the channel off the Calstock Boatyard on the port hand bank; these are sometimes available for visitors drawing up to 1m but it is best to phone ahead to enquire (Tel: 01822 832502). Deeper draught boats may be able to dry out alongside their small quay, where diesel is available.

Because of the moorings, the poor holding, and the narrow channel used by the pleasure boats, anchoring is not recommended downstream of the viaduct, and the only feasible place for a visitor to stop is beyond it, just upstream of the pleasure boat landing pontoon, where there is a pool with just over 2m LAT.

Once the busiest port in the upper Tamar, looking at Calstock today it all seems inconceivable; the quays where ships lay two or three abreast have mostly crumbled into disrepair, and the attractive cluster of cottages and small Georgian houses clinging to the steep roads up the hillside betray little evidence of their busy past. On the opposite shore, now lost in the reeds and sedge, the famous

Cargreen; looking upriver to Weir Quay

shipyard of James Goss was building large wooden vessels as late as 1909 when the last, the ketch *Garlandstone* was launched, and she is still afloat today. If you need proof, or an excuse for a pint, go and take a look at the photographs in the pub.

Most normal facilities are available, including two pubs and a restaurant; petrol at the garage, and a branch of Lloyds bank open Mondays only, 1000-1230. There are also trains to Plymouth.

Beyond Calstock the river is becomes much narrower, very shallow and tortuous, and is best explored further by only dinghy, or small shallow draft boats. Morwellham, just over two miles upstream, is a monument to the industrial past of the Tamar, a former copper port restored by the Dartington Amenity Trust. Open daily during the season as a tourist attraction, it is even possible to take a train ride deep into one of the old copper mines.

PORT GUIDE: PLYMOUTH
AREA TELEPHONE CODE: 01752

Harbour Master: Commander R G Allen, RN, Queen's Harbour Master, HM Naval Base, Plymouth. (Tel: 552047) Deputy QHM, Longroom, (Tel: 663225). Cattewater Harbour Master, Commander Anthony Dyer, 2, The Barbican, Plymouth PL1 2LR, (Tel: 665934). Sutton Harbour Master, Mr P Marshall, Harbour Office, Guy's Quay, Sutton Harbour, Plymouth. (Tel: 664186)
VHF: Channel 16, working 14, call sign *Longroom Port Control* (24 hours). *Cattewater Harbour Office* 16,12,14 and 80, (Mon-Fri 0900-1700). *Sutton Harbour Radio* 16,12,80 and 37, (24 hours). See 'marinas' below
Mail Drop: Marinas. RWYC and RPCYC
Emergency Services: Lifeboat. Brixham Coastguard
Anchorages: Cawsand Bay. Off Jennycliff. Barn Pool. Drake's Island. St German's and Tamar Rivers
Moorings: RWYC, RPCYC. Weir Quay Boatyard (Tel: 01822 840474). Calstock Boatyard, (Tel: 01822 832502)
Dinghy landings: RPCYC. Mayflower SC
Water Taxi: QAB to Mayflower Steps. Mayflower Steps to Turnchapel.(Clovelly Bay)
Marinas: Mayflower International Marina, Ocean Quay, Richmond Walk, Plymouth PL1 4LS, (Tel: 556633), 270 berths, 30+ visitors, call sign *Mayflower Marina* VHF channels 80 and 37 (24 hours). **Queen Anne's Battery Marina**, Plymouth PL4 0LP, (Tel: 671142) 240 berths, 60+ visitors, call sign *QAB* VHF channels 80 and 37 (24 hours). **Sutton Harbour Marina**, Sutton Harbour, Plymouth, PL4 0ES, (Tel: 664186). 310 berths, 30 visitors. Approached through free lock (access 24 hours), call sign *Sutton Harbour Radio* VHF channels 16,12,80 and 37 (24 hours). **Clovelly Bay Marina**, Turnchapel, Plymouth PL9 9TF, (Tel: 404231)190 berths, 20 visitors, call sign *Clovelly Bay* VHF channel 37 and 80 (24 hours)
Charges: None for anchoring. RWYC mooring £7.50 p/n. RPCYC £5 p/n. Weir Quay £5 p/n. Overnight charge for 30 footer (inc VAT): Mayflower International Marina £14.85. Queen Anne's Battery Marina £15.86. Sutton Harbour Marina £15. Clovelly Bay Marina £12
Phones: At all marinas and yacht clubs. Cawsand, Cremyll, St Germans, Cargreen, Calstock
Doctor/dentist: Ask at marinas
Hospital: Derriford, (Tel: 777111)
Churches: All denominations
Local weather forecast: At marinas
Fuel: Mayflower International Marina, diesel and petrol 24 hours. Queen Anne's Battery Marina, diesel and petrol,

0830-1830. Sutton Harbour Marina, diesel and petrol, 0830-1830. Clovelly Bay Marina, diesel available in cans 24 hours, alongside around HW only. Weir Quay and Calstock Boatyard, diesel only
Paraffin: At QAB and Mayflower marinas
Gas: Calor/Gaz at all marinas, 24 hours at Mayflower Marina
Water: At all marinas. In cans, Cawsand, Multihull Centre, Millbrook. St German's Quay, Cargreen, Weir Quay, Calstock
Tourist information office: Barbican, opposite old Fishmarket
Banks/cashpoints: All main banks in Plymouth city centre, all with cashpoints
Post Offices: Barbican. Plymouth City centre. Cawsand. Millbrook. St German's. Cargreen. Calstock
Rubbish: Disposal facilities at all marinas
Showers/toilets: At all marinas. RWYC. RPCYC
Launderettes: At all marinas
Provisions: Everything obtainable. EC Wednesday but most shops remain open and many shops also open on Sundays
Chandler: Cloads, The Barbican (Tel: 663722). Yacht Parts Plymouth, Queen Anne's Battery (Tel: 252489). Sutton Marine, Sutton Pier (Tel: 662129) Ocean Marine Services, Mayflower Marina (Tel: 500977) and also QAB (Tel: 222550). Vosper Marine, Sutton Harbour (Tel: 228569). A E Monson, Admiralty Chart Agents and bonded stores. The Sea Chest, Queen Anne's Battery, nautical bookshop/Admiralty Chart Agent, (Tel: 222012)
Repairs: Mashford Bros, Cremyll (Tel: 822232). Portway Yacht Care, Mayflower Marina (Tel: 606999). Harbour Marine, Sutton Harbour (Tel: 666330). Danvic Boatyard Ltd, Queen Anne's Battery (Tel: 268677). K R Skentleberry & Son, Laira Bridge Boatyard (Tel: 402385). A Blagdon, Richmond Walk (Tel: 561830). A S Blagdon & Sons, Embankment Road (Tel: 228155). Ron Greet, Turnchapel Marine (Tel: 402969). Multihull centre, Millbrook (Tel: 823900)
Marine engineers: Harbour Marine (Tel: 666330). Marineserv (24hours) (Tel: 346956). M&G Marine, Mayflower Marina (Tel: 862277). Ask at marinas or boatyards
Electronic engineers: Sutton Marine (Tel: 662129). Tolley Marine (Tel: 222530). Ocean Marine Services (Tel: 500977). Ask at marinas
Sailmakers: Clements (Tel: 562465). Westaway Sails, Ivybridge (Tel: 892560). Ask at marinas
Riggers: AB Yacht Riggers (Tel: 226609). Peter Lucas Rigging, (Tel: 551099). Sutton Rigging (Tel: 269756)
Transport: Main line trains to London and the north. Good road connections to M5. Plymouth airport, flights to London/Scotland/Eire. Ferries to Roscoff and Santander
Car Hire: Hertz (Tel: 705819). Avis (Tel: 221550) or ask at marinas
Taxi: (Tel: 669999)
Car Parking: All marinas have customer parking
Yacht Clubs: Royal Western Yacht Club of England, Queen Anne's Battery, Plymouth PL4 0LP (Tel: 660077). Royal Plymouth Corinthian Yacht Club, Madeira Road, The Hoe, Plymouth PL1 2NY (Tel: 664327). Mayflower Sailing Club, Phoenix Wharf, Plymouth (Tel: 662526). Saltash Sailing Club, Waterside, Saltash (Tel: 845988)
Eating out: Vast choice from Indian to Greek, Spanish to Chinese. Restaurants, bistros, pub food, fish and chips
Things to do: Elizabethan House, Plymouth Museum in old merchant's house, Royal Citadel, Barbican, Smeaton's Tower, Plymouth Dome and Pavilions, audio/visual history of Plymouth, The Hoe. Large shopping centre
Regattas/special events: Port of Plymouth Regatta, end July. Venue for many special events, including dinghy championships, and start/finish of major offshore races, including Fastnet, Transatlantic and Round Britain

Passages
RAME HEAD TO THE MANACLES

Favourable tidal streams

RAME HEAD	Bound West: 2 hours before HW Dover
	Bound East: 4 hours after HW Dover
DODMAN	Bound West: 3 hours before HW Dover
POINT	Bound East: 3 hours after HW Dover

Passages – Rame Head to The Manacles
Charts for this section of coast are:
BA: 1267 Falmouth to Plymouth 148, Dodman Point to Looe Bay (including harbour plan of Polperro). 777 Land's End to Falmouth. 3l Harbours on the south coast of Cornwall (Fowey, Charlestown, Par). 147 Plans on the south coast of Cornwall (Helford River, Looe, Mevagissey)
Imray: C6 Start Point to Lizard
Stanford: 13, Start Point to Land's End and Padstow
French: 4812. Du Cap Lizard à Start Pt

RAME to MANACLES

Beacon	Range (M)	Frequency (kHz)	Call Sign
Marine beacon			
Lizard	70	284.5	LZ
Aero beacon			
Plymouth	20	396.5	PY

Coastal Radio Pendennis Radio, working channel 62

West of Plymouth and the Yealm, Fowey is the next probable destination on a coasthopping cruise and at over 20 miles away, it is one of the longest legs. Although no great distance in itself, yet again the odds are very much in favour of winds forward of the beam and, once past Rame Head, the tidal streams are weak inshore and of no great assistance. Three miles south of Looe the streams rotate in a clockwise direction, attaining a maximum at Springs of about one knot, east by north HW Dover –0520, but no more than ½ knot west by south HW Dover +0035. Somehow, this passage always seems to take longer than anticipated.

In the far distance, given reasonable visibility, twenty-three miles WSW of Rame Head is the long, flat topped headland of the Dodman, with its distinctively rounded end. However, immediately to the north-west of Rame Head, Whitesand Bay falls back, a succession of sandy beaches and rocky outcrops, backed by a continuous line of steep and broken grassy cliffs, between 30m and 76m high. The direct course to Fowey keeps you a good couple of miles offshore, which is no great disadvantage as this is one of the less inspiring stretches of coast. Rocks extend a cable immediately to the west of Rame Head, and north-west of the point is a dangerous wreck awash at LWS marked by a red can. Following the sweep of the bay inshore there are rifle ranges by the old fort at Tregantle, where there are a number of flagstaffs along the cliffs, and if the red flags are flying you should keep well clear.

Rame Head looking westwards, with Downend Point in the distance

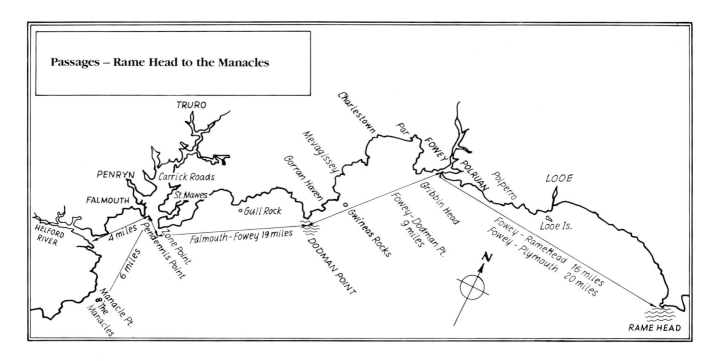

Passages – Rame Head to the Manacles

Sounding like a perfect setting for an Enid Blyton children's adventure, **Portwrinkle**, almost two miles west of Tregantle, is a very tiny boat harbour, which dries completely and is fringed by extensive rocky ledges. There are a few houses nearby, and the village of Crafthole can be seen on the skyline. From here onwards, the coast continues in a long, high sweep with extensive rocky ledges and hazards extending up to two cables offshore, including the Longstone, a prominent rock 18m high. At **Downderry**, which straggles along the cliffs, precipitous paths lead down to a beach, and to the south there are the Sherbeterry Rocks, a large area of shallows extending just over two miles offshore, with least depths of 4.9m, which can produce an area of rough water particularly in onshore winds when this whole stretch of coast should be avoided, and a good offing maintained.

Looe Island, a rounded hump, 44m high, lies close to the shore, almost linked to the mainland at LW, a perfect natural daymark for the drying fishing port of **Looe** just to the east, which is mostly commercial with very limited facilities for yachts, and is dangerous to approach in south or south-easterly winds of any strength and ebb tide as it

has a long and very narrow entrance. In offshore winds, however, there is a good anchorage just off the harbour. Rounding Looe Island, the Rennies, a group of drying rocks extend three cables south-east from the island, and at times, in fresh wind against tide a small race will be found up to one mile from the shore. From offshore, a very tall TV mast, 586m high, can be seen inland, 10 miles due north of Looe Island, displaying a number of fixed red lights at night.

Beyond Looe the coast becomes more interesting again. There is a measured mile (1852.9m to be precise!) just west of the island, the transits formed by two white beacons with a vertical black stripe at each end, running as far as the approach to Talland Bay. This is an attractive sandy beach backed by trees and fields popular with holidaymakers and its end is formed by the 100m high headland of Downend Point, which has a large granite war memorial near the summit. Downend Shoals, ½ mile due south, have a least depth of 2.6m LAT and should be passed well to seaward if there is any sea running. If bound for **Polperro**, the small fishing village harbour just over ½ mile west, it is best not to turn inshore until the white beacon

The coastline west of Downderry, Looe Island on right

Polperrro lies in a narrow cleft in the cliffs

on Spy Glass Point (Iso WR 6s 7M) just east of the harbour is bearing north. Approaching from the west, a prominent TV mast will be seen high on the cliffs just before the village opens. Like Looe, Polperro is not a place much frequented by yachts, and although space is very restricted, there is usually room for a couple of visitors to dry alongside the quay. The alternative is a mooring or anchoring just south of the harbour. In bad weather a storm gate closes the harbour completely.

From Polperro to Fowey, in good weather, the coast is very attractive with impressive cliffs up to 91m in height occasionally broken by steep green and grassy coombes and gullies running down to small coves. The water is deep to within two cables of the shore and there are no hazards beyond that, except the insidious **Udder Rock**, which dries 0.6m, two miles west of Polperro and ½ mile offshore, which is marked by an unlit south cardinal buoy, YB. A white beacon on the cliffs, and a white mark on a rock on the shore also provide a transit for this hazard, bearing 020°T, and a prominent white mark on the western side of Lantic Bay just open of steep Pencarrow Head, 135m high, provides a cross bearing of 283°T. At night, if you are in the red sector of Fowey light you will pass Udder Rock to the south but, better still, keep further to seaward and open the white sector. Generally, if making the passage from Plymouth to Fowey at night, the dominant feature is the Eddystone to the south-east, (Fl (2) 10s 24M) but you will be on the 13M limit of the fixed red sector visible through 112-129°T. Rame Head is unlit, and beyond it the few shore lights are Looe harbour entrance (Occ WR 3s 8M), an (Iso WR 3s 7M) light east of Polperro Harbour and Fowey (L Fl WR 5 sec, W 11M, R 9M).

In daylight, although completely hidden from the east, the entrance to Fowey harbour is not difficult to find, thanks to the huge red and white horizontally striped daymark, a square pillar 104m high built by Trinity House' in 1832 on Gribbin Head which extends as a long promontory to the south-west of the river mouth. Lantic Bay, immediately west of Pencarrow Head, is a mile east of the entrance, backed by National Trust land, and the beach, though mostly covered at HW, is a beautiful stretch of clean sand when the tide is lower, with a good anchorage off it in northerly winds.

Beyond it there is a conspicuous white house high on the cliffs, and a ruined round tower. As the headland draws abeam, Punch Cross, a white cross on a large rock, will come into view to starboard, with a Coastguard lookout and houses along the cliff above, and the river mouth and town of Fowey will emerge, with high ground on either side. Deep and free from hazards, it can be approached in any weather, although obviously strong southerly winds and ebb tide will produce very rough conditions and breaking seas in the entrance. As well as being a popular yacht harbour, this is also a very busy commercial port, exporting large quantities of china clay, so be ready for surprise encounters with sizeable ships in the entrance which is only a cable wide at its narrowest point. They, of course, have priority, so keep well clear.

The passage from Fowey to Falmouth is about 20 miles, and the tide is once again a definite factor to consider for rounding the Dodman. The Lizard and Start Point both instantly evoke races, but somehow the Dodman seems to elude such an association, which is strange, as there can often be quite an unpleasant amount of disturbance in its vicinity. Close to the point streams run at nearly two knots at Springs, and with wind against tide, the uneven depths and shoals of the Bellows and Field can produce a small but very unpleasant race extending a good mile offshore. If a westerly wind of any strength is prevailing ideally try to round the Point at slack water just before the main ebb begins to the south-west, about three hours before HW Dover, which means leaving Fowey two hours after local HW (Dover –0600). The whole promontory of the Dodman provides a considerable lee in westerly weather, and it should be remembered that a fresher wind and larger seas are likely to be encountered once past the point, particularly in a south-westerly when it can be a long ten mile beat to windward to reach Falmouth, along an exposed stretch of coast that provides no shelter in between. The only hazards between Fowey and the Dodman are both well marked; Cannis Rock, drying 4.3m, inshore of the 'Cannis Rock' south cardinal buoy YB (Q6 + L Fl 15 sec) positioned ½ mile south-east of Gribbin Head, and also covered by the western red sector of Fowey lighthouse, and the Gwineas Rocks, the largest of which dries 8m just

Cannis Rock buoy and Gribbin Head daymark

over two miles north-east of Dodman Point, with 'Gwineas' east cardinal buoy BYB (Q (3) 10 sec) two cables to the south-east.

Clearing Gribbin Head, St Austell Bay opens to the north-west, and the coast running away towards the Dodman, nine miles south-west of Fowey provides a good sheltered area of water in westerly winds, an attractive miniature cruising ground in its own right. Beyond it, the land rises, the distant sprawl of houses around **St Austell** culminating in the high, jagged skyline of the 'Cornish Alps' – the huge spoil heaps surrounding the extensive china clay workings of this important local industry. From offshore, when they catch the sun, the similarity to a range of snow covered mountains is an undeniably remarkable, and possibly confusing sight.

The china clay port of **Par** is commercial and of no interest to visitors, but does provide another very distinctive landmark on the northern shore of the bay, four large

Rame Head to The Manacles

Dodman Point from the north-east

chimneys that are even more conspicuous when belching out white smoke.

Your course will take you well away from the land, closing it east of the Gwineas Buoy, but the town and white lighthouse of **Mevagissey** are easy to spot on the western shore, a V-shaped gap in the cliffs, surmounted by houses on both sides. **Gorran Haven**, a drying small boat harbour lies due west of Gwineas rocks. As always, particularly near the rocks, keep a good lookout for pot buoys.

Close to, the **Dodman** is impressive, a rounded bluff, 111m high, flat topped with steep cliffs along the eastern side but with a more sloping western side, a lovely stretch of National Trust property covered in gorse, ferns and grass. High on the south-western tip there is a large white cross erected in 1896 by a local vicar.

On a fine calm day, it is quite feasible to pass within a cable of the foot of the cliffs, as the water is deep and unimpeded; in any sort of sea or weather, stand well out, up to two miles in fresh conditions. A notorious headland for shipwrecks, particularly in fog, it is strange that no light or fog signal was ever established there, particularly as both St Anthony light (Occ WR 15s 22M), and the Lizard light (Fl 3s 29M) are obscured from the Dodman. Apart from the lights of villages ashore there are no aids to navigation at night until St Anthony light appears on a north-westerly bearing, although sometimes its loom will be seen sooner. There is also a noticeable set into the bays between

the Dodman and Falmouth; in poor visibility or at night give the coast a good berth.

Once past the Dodman, Veryan Bay opens to the north, an attractive but uncompromising cliff-lined stretch of coast, broken at intervals by small sandy coves.

The western extremity of Veryan Bay, Nare Head, five miles distant, looks similar to the Dodman, but has the distinctive triangle of Gull Rock, an island 38m high ½ mile to the east. Although Veryan Bay is almost free from offshore dangers, with no rocks more than two cables from the shore, there is one notable exception, Lath Rock, least depth 2.1m LAT, almost midway across the bay, fortunately just inside the direct line from the Dodman to Gull Rock.

Gull Rock is a jagged pyramid, with sparse grass around its whitened summit, clear evidence of the many seabirds that nest there and give it its name. In fair weather it is possible to pass between the rock and the steep cliffs at Nare Head, 80m high, but the Whelps, a reef with a number of drying rocks, mostly 4.6m, extends ½ mile to the south-west of Gull Rock.

Gerrans Bay, much of which is surrounded by National Trust land, is of a similar aspect to Veryan, although the cliffs along its western side become less precipitous. Probably the best of all the anchorages along this section of coast is off **Porthscatho** which does give shelter from the west, a small fishing village now popular with holiday-makers. There is a small pier on its southern side which gives protection to the drying foreshore, with a reasonable anchorage just to the north-east. Basic provisions, Post Office and several pubs ashore make this a popular day sail from Falmouth.

Gerrans Bay has seen many shipwrecks in its time, with vessels mistaking it for the entrance to Falmouth, but the most spectacular in recent years was the capsize of pop star Simon Le Bon's maxi, *Drum* after she lost her keel in gale force conditions during the 1985 Fastnet.

The Bizzies, least depth 4.2m LAT, lie almost on the line from Gull Rock to Porthmellin Head, and should be avoided in fresh winds and ground swell which can create quite an area of overfalls around them.

Beyond Gerrans Bay the rounded profile of Zone Point continues to hide the elusive St Anthony light, and a certain nagging doubt can tend to creep in at this stage of the

Dodman Point from the south. Note Gribbin Head daymark in the distance

passage. There used to be a conveniently conspicuous row of coastguard cottages high on Zone Point which have since been demolished by the National Trust. It is surprising that no daymark was ever established here as the entrance is not easy to locate from the east until, at last, the white lantern peeps into view from behind the headland. Just over a mile south of the light, Old Wall, a rocky shoal with least depth of 7m rises steeply from the sea bed and can produce an area of rough water in strong southerly winds. In fine weather it is a popular fishing area and easy to spot from the number of angling boats in its vicinity.

The entrance to the River Fal is a mile wide and safe to enter in any conditions, the only hazard being Black Rock right in the middle, marked by a conspicuous but unlit beacon which can be passed either side. In strong southerly winds, with an ebb tide out of the estuary, rough seas will, however, be encountered in the approach and entrance. The ebb, which can be up to two knots at Springs, if fresh water is running down the river, begins at HW Dover –0605.

It is just over five miles from St Anthony Head to Manacle Point, away to the south-west, and between them Falmouth Bay is a fine natural roadstead, well sheltered from the north to south-west, but open to the east and south, much used by large vessels as an anchorage, and increasingly so for offshore bunkering.

From Pendennis Head, neatly crowned with its castle, the hotels and beaches of the Falmouth seafront form a broad sweep to the west, and the large cream coloured Falmouth Hotel at the eastern end is particularly prominent. Depths reduce gradually towards the shore, which is fringed with rocky ledges, but there are no dangers further than a cable from it, except large numbers of poorly marked pot buoys. In offshore winds it is possible to anchor off Swanpool Beach just north of the prominent and wooded Pennance Point; a mile to the south Maenporth is another popular cove, and during the summer a large inflatable racing mark is usually anchored to seaward.

From Falmouth, the entrance to the **Helford River** is not easy to distinguish; the various headlands of similar shape blending together. Low cliffs run between Maenporth and Rosemullion Head, which is flat topped and rounded, covered in gorse and thick bushes. Between it and Mawnan Shear, there are steep grassy cliffs, with a dense clump of woods, and a conspicuous white house at its eastern edge. The **Gedges** rocks, which dry 1.4m LAT, lie three cables ESE and are marked to seaward by the green conical 'August Rock' buoy, during the summer months only. This is the only real hazard in the approach to the Helford River, and once past the buoy the entrance opens clearly. It is exposed to the east, when the shallowing water produces particularly steep short seas, and being unlit should not be attempted at night without local knowledge.

Gillan Creek, due south of the Gedges, is an opening in the southern approach to the Helford, an attractive, but mostly drying inlet between Dennis Head, which is grassy and 43m high, and the much lower promontory of Nare Point with an old square Coastguard lookout on the end. Beware Car Croc, a rock just awash at LW in the entrance of Gillan Creek, marked with an east cardinal buoy BYB which should be left to starboard when entering. There are also rocky ledges extending a cable to seaward of Nare Point.

Proceeding south from Falmouth or the Helford, the extensive rocky nightmare of the Manacles involves a detour to the east. The yellow 'Helston' buoy (Fl Y 2.5s), known locally as the three mile buoy, lies just over three miles south of St Anthony, and almost the same distance east of the entrance to the Helford River. This part of the bay is much used for search and rescue exercises from RNAS Culdrose, and if you see Admiralty vessels and helicopters hovering in the vicinity keep well clear.

Manacle Point is a rather untidy looking headland, badly scarred by extensive old quarry workings, and just to the north, by the small cove of Porthoustock, are the unsightly remains of the huge stone loading chutes on either side of the bay. Jagged pinnacles of rock extend from the point, which continue to form the reefs offshore, marked by the 'Manacle' buoy, east cardinal (Q (3) 10 sec), 1½ miles to the east, which always tends to appear further out to sea than you anticipate, particularly when approaching from the south. The Manacles are undoubtedly one of the most treacherous hazards along the Cornish coast. Right in the approach to a busy port, it was inevitable that this area of half-tide rocks and strong currents should claim so many ships over the years. Their sinister name actually derives from the Cornish 'maen eglos', meaning Church stones, for the spire of St Keverne Church is prominent inland.

Close to the Manacles the tidal streams run at up to two knots at Springs; if bound round the Lizard, aim to leave Falmouth or the Helford about three hours after local HW to gain the best advantage. There is an inshore passage through the Manacles regularly used by local fishing boats but do not be tempted to follow them as it is very narrow in places with unpredictable eddies and currents exceeding three knots. Do not be too amazed, either, if you see a reasonable sized coaster seemingly emerging from among the rocks. Incredibly, they regularly load stone at Dean Quarry alongside the cliffs just south of the Manacles.

At night, approaching from the south, the Manacles lie within the red sector of St Anthony light, 004°-022°T.

Before radar a very large percentage of the wrecks along this section of coast occurred in calms and fog rather than extremes of weather, and in poor visibility it can be very difficult, as a number of the headlands and bays have a similar appearance and audible aids are few and far between; Eddystone Light (Horn (3) 60s), Nailzee Point, Looe (Siren (2) 30s occasnl), Cannis Rock buoy (Bell) – note there is no other fog signal at Fowey – Mevagissey (Dia 30s), Gwineas buoy (Bell), St Anthony Head (Horn 30s), Manacles Buoy (Bell). Remember that there is generally a northerly set into the bays, and although the 10m sounding line provides a good indication of relative position clearing most of the hazards close to the shore, such as the Dodman, this is little more than two cables off. If in doubt, err to seawards, and do not follow local fishing boats, particularly close to the Manacles – there is no telling where they might be going.

LOOE AND POLPERRO

Tides: HW Dover –0545. Range: MHWS 5.4m – MHWN 4.2m MLWN 2.0m – MLWS 0.6m. Looe harbour, strong streams at springs
Charts: BA: 147,148. Stanford: 13. Imray: C6
Waypoints: Looe Banjo Pier Head 50°21′02N. 04°27′00W. Polperro West Pier Head 50°19′84N. 04°30′92W
Hazards: Looe: Looe Island and Rennies rocks to south (all unlit)
Polperro: The Rannys and Polca Rock in approaches (both unlit). Both harbours dry to entrance, and are dangerous to approach in onshore wind and sea. Polperro harbour mouth closed in bad weather. Busy fishing harbours, keep clear of local boats. Beware pot and net buoys in approaches

Midway between Plymouth and the next popular cruising haunt of Fowey lie the small harbours of Looe and Polperro. Working fishing ports, with restricted space, both dry almost completely at LW. Boats unable to take the ground comfortably or lie alongside can anchor off, or in the case of Polperro, use the moorings just outside the harbour. With narrow entrances, neither harbour should be approached in onshore winds of any strength.

Approaches

Looe lies nine miles west of Rame Head at the mouth of the River Looe, which divides the town into East and West Looe, linked by a bridge with 3m clearance across the upper end of the harbour. From offshore, the entrance is easily located, lying to the east of the large rounded lump of Looe Island, 44m at its highest point, half a mile offshore. This is privately owned, the residence of two remarkable sisters, Evelyn and Babs Atkins who threw up a civilised existence in suburban Epsom and moved on to the island in 1965, a tale related in the book 'We Bought an Island'. Tales of contraband and smuggling inevitably abound but the most bizarre episode in the island's history had to be during the last war when it was bombed by an over zealous German pilot who mistook it for a large warship. Today it is open to the public, and a small landing fee is charged, the proceeds going towards its upkeep. It is also known locally as St George's Island and, if approaching from the west, keep a good half mile to the south to avoid the drying Rennies Rocks (4.5m). South of them, in certain combinations of wind and tide quite a brisk area of overfalls can form.

Do not attempt to pass between the island and the mainland, which is rock strewn and shallow and, ideally, approach Looe from the south-east, leaving the island a good two to three cables on the port hand as you close the land and head directly for the end of the Banjo Pier – its shape gives it the name – which forms the eastern side of the harbour mouth. Midmain is an isolated rock south-west of the entrance marked by a lighted beacon with east cardinal topmark (Q (3) 10s), and on the rocks to the north there are two red beacons closer to the harbour.

If waiting for the tide a good anchorage lies just east of the pier head, 2m LAT. The entrance is long and little more than 50m at the narrowest point, drying right to the mouth at Springs, and accessible after half flood for boats of average draught. It is flanked by high cliffs along the western side, with rocky ledges at their foot. Only enter under power, keeping close to the Banjo Pier. The tide runs strongly, up to 5 knots on the ebb at Springs, and careful allowance should be made for this when manoeuvring. Beware, too, of the numerous self drive motorboats – always an unpredictable hazard – the fishing boats entering and leaving, and the small passenger ferry at the southern end of the harbour. Although the pier head is lit, (sectored Occ WR 3s) – the white light covering the safe approach – entry at night is not recommended due to the narrow entrance and absence of lights within. It is far safer to anchor off and enter in daylight; because of the fishing boat movements, a riding light is essential.

Berthing

The quays on the east side of the harbour are reserved for the fishing fleet, and further upstream the Harbour Master's Office is in the fishmarket (Harbour Master Mr E T H Webb, Tel: 01503 262839). There is no VHF watch. Below

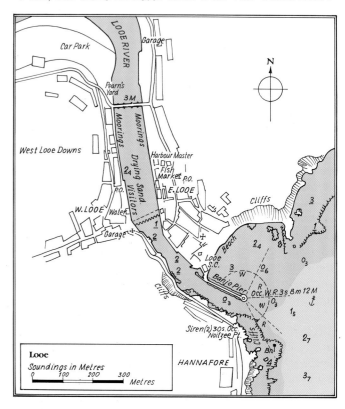

Looe
Soundings in Metres

There is just one drying visitors' berth at Looe on the west side of the busy harbour

the bridge a number of local boats lie on trot moorings, all of which dry, and the only berth available is clearly marked with a yellow sign 'Visitors Berth Only' at the southern end of the West Looe quay, just above the ferry steps. You should report to the Harbour Master on arrival; the charge per night is 30p per foot, plus VAT. The quay is lined with wooden piles and a fender board will be very handy; you will ground about four hours after HW, drying out on a firm sand and gravel bottom. Anchoring is prohibited throughout the harbour because of the moorings and underwater pipes and power cables.

Facilities

A picturesquely situated little town, Looe clings to the steep hills overlooking the river and during the season it pays the inevitable price and is absolutely inundated with visitors, the crowds thronging through narrow streets that were never designed for such an influx. In East Looe gift shops abound, but West Looe retains more of the original feel of this old fishing community with thick-walled pastel painted cottages clustered around a network of narrow alleyways and cobbled yards.

Facilities are good, with plenty of shops, several open Sunday morning, a launderette in East Looe, numerous restaurants, cafes and pubs. Fresh water is available from taps on both quays, and there is a convenient shower/toilet adjacent to the visitors' berth. Diesel is available alongside from 'Looe Fish'; petrol in cans from the local garage. There is a Post Office in east and west Looe, branches of Lloyds, Midland and Barclays banks, a chandler, and two boatyards upstream of the bridge both with slipping facilities. There are also several marine engineers. Looe Sailing Club, in Buller Street, East Looe has a licensed bar and showers are available. A railway branch line connects with the main line at Liskeard.

Like many other Cornish fishing ports Looe has had its ups and downs, suffering some very lean times after the collapse of the pilchard industry in the 1930s, and the mackerel fishery in the 1970s. Since then the fleet of medium sized boats has been expanding, mostly potting, long-lining, trawling and scalloping. The holiday industry has in many ways helped to sustain the commercial fishermen through some of the harder times, as this has long been a popular sea angling port and home of the Shark Fishing Club of Great Britain and, though nowhere in the league of 'Jaws', a surprising number of Porbeagle and Blue sharks are caught in the warm waters off Cornwall during the summer. Definitely not man-eaters, they are rarely found close to the shore.

At the turn of the century a fleet of over 50 large luggers still worked from Looe, decked boats up to fifty feet, their transom sterns distinguishing them from the double ended luggers of the west Cornish ports. The pilchards, small fish similar to herring, were traditionally caught in long seine nets during the summer months, with open boats working from coves and beaches, but as the shoals moved further offshore, the larger boats were built and long drift nets began to replace the seine.

However, of all the Looe luggers, probably the best known, and certainly the most travelled, has to be the *Lily*, built by Ferris in 1896. Her fishing days over sometime in the 1930s, a young, enthusiastic couple spotted her in nearby Polperro.

'Anne stood by the helm. "I think," she said, "we've found our ship." We scrambled aboard and went below. The fishwell took up the whole of the middle of the ship. We pulled up the floorboards and looked into the bilge. There were firebars, cannonballs, cogwheels and shingle. She was dry as a bone but smelt to heaven of fish and tar. In the fo'c'sle was a bogie stove eaten with rust, and a single locker on the starboard side. Aft was a single-cylinder Kelvin. Anne's face was flushed with excitement. "If she only costs us twenty-five pounds, we can afford to have her converted into just what we want. It'll be like buying a new boat." I hadn't the heart to tell her that a surveyor might find she was rotten.'

She turned out to be very sound indeed, and thus a humble Looe lugger was revived to become Peter and Anne Pye's famous gaff cutter *Moonraker of Fowey*. While her sister ships were towed to a final, rotting resting place up the Looe River, for it was considered unlucky to break them up, *'Moonie'* suffered no such ignomony, ranging far and wide from the Baltic to Tahiti, and Alaska to Brazil, her voyages delightfully described in 'Red Mains'l', 'The Sea is for Sailing', and 'A Sail in a Forest'.

POLPERRO

There are no such bargains to be found in Polperro today, and the people who throng there are not searching for boats to buy, merely postcards, cream teas and souvenirs. It is not, however, a particularly recent phenomenon, as a guide book written in the 1920s reveals. . .

'. . . a human bees' nest stowed away in a cranny of rocks. Its industries are four: the catching of fish, the painting of pictures, handicrafts and – latterly – the entertainment of visitors, for it is probably the most charmingly unexpected village in all England. . .'

The pressure of visitors in recent years has resulted in the welcome banning of all cars from Polperro during the day, with a large car park on the outskirts. The ten minute walk down the hill comes as a considerable shock to many of today's car-bound explorers unused to such inconvenience and exertion; with great relief, they soon spot the horse drawn bus.

Rame Head to The Manacles

Approaches

This undeniably attractive harbour is set in a narrow cleft among the cliffs just under three miles west of Looe Island, and five miles east of Fowey. It is not, at first, the easiest place to find and no attempt should be made to approach it in any southerly wind and seas, particularly south-east. Not only are the approaches dangerous in these conditions, it is quite likely that the protective steel gate across the entrance will be closed.

Approaching from the east, once past Looe Island, a pair of transit beacons will be seen on the mainland marking the beginning of a measured mile; the second set of beacons marking its end lie just east of Talland Bay, with Downend Point at its western end. From south of the point a V-shaped cleft in the high cliffs will begin to open, and houses of Polperro will be seen. Steer to within two cables of the inlet, and Peak Rock will be seen on the western side, pyramid shaped, topped with jagged boulders, with Spyglass Point on the eastern side where the harbour light (Iso WR 3s) is located. The main hazard is Polca Rock, (1.2m) about 200m south-west of the point. The Rannys, two rocks drying 0.8m LAT, will usually be seen awash just off Peak Rock; proceed with caution, passing between these and the hidden Polca, before turning into the inlet and heading for the east pier end. If the tide is low or you do not wish to dry out there are six buoys in the approach. If you intend to stay overnight it is best to moor fore and aft (N/S) between them to avoid swinging into the fairway to the harbour; alternatively, anchor to seaward clear of them in 3m – 4m LAT with a riding light.

Moorings

The inner harbour dries to just beyond its mouth at Springs, and can be entered after half flood. Fishing boats use the inside of the western pier, and the rest of the harbour has many fore and aft small boat moorings. Berthing for visitors is very limited, with space for a couple of boats alongside the east quay wall immediately inside the entrance. Charges for the use of the moorings or quay are around 27p a foot per night plus VAT and Chris Curtis, the Harbour Master, will appear at sometime to collect it. His office is in the small fish market on the western side of the harbour (Tel: 01503 72809) but he is also a full-time fisherman; when he is at sea he can usually be contacted on

There is very limited space in Polperro. Visitors can anchor off, or lie to a mooring

VHF channel 10 call sign *Polperro Harbour Master* should the need arise.

Facilities

In spite of its popularity the village has, fortunately not been over developed, and is carefully administered by the Polperro Harbour Trustees, an entirely voluntary organisation, with the declared intention of maintaining its unique character. The tiny narrow streets, and houses, overhanging the harbour and the diminutive River Pol, apart from the smart paint and good state of repair, have probably outwardly changed little since John Wesley was forced to change his lodgings in 1762 because the room beneath him was *'filled with pilchards and conger eels, the perfume being too potent . . .'* Prior to the advent of the tourist industry, fishing, and smuggling were the main activities, the rabbit warren of alleyways ideal for evading the Excise men. Special boats were built for the trade, and one, *Unity*, was reputed to have made 500 successful trips, crossing to France in just over eight hours with a fair wind. Today this colourful past is well documented in the 'Polperro Heritage Centre of Smuggling and Fishing' in the old pilchard factory on the east side of the harbour.

In spite of the crowds, it is still an intriguing place to explore, particularly in the evening when the coaches and day trippers have all gone on their way. Facilities are limited to provisions, Post Office and branches of Barclays and Midland banks, open 1000-1430, May-September. There are several restaurants, and four pubs, the nearest fuel in cans is from a garage about a mile distant, although the local fishermen will always help out with a gallon or two of diesel if you're desperate. There are water taps on both quays.

In November 1824, three houses, the inner and outer piers and fifty boats were destroyed in the worst south-easterly storm ever experienced. On a balmy summer's evening, fortunately, Polperro is a very different place, the cheerful red, green and yellow of the boats and the subtle backdrop of white, pink and slate hung cottages shimmering on the water at the height of the tide, while lingering holidaymakers bask on the warm, weathered stone of the harbour wall, eking out the very last of the sun's fading glow as the valley sinks into shadow.

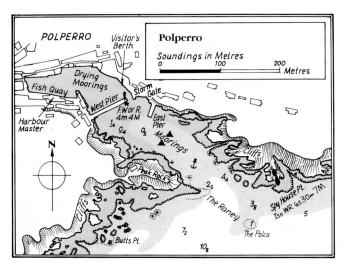

FOWEY

Tides:	HW Dover –0550. Range: MHWS 5.4m – MHWN 4.3m. MLWN 2.0m – MLWS 0.6m
Charts:	BA: 31. Stanford: 13. Imray: Y52
Waypoint:	Cannis Rock buoy 50°18′35N. 04°39′88W
Hazards:	Udder Rock 3M to east, Punch Cross (both unlit). Cannis Rock (lit). Busy commercial port, beware shipping in narrow entrance. Upper reaches dry; main harbour dredged to 7m. Fluky winds within harbour

'*Fowey is the harbour of harbours, the last port town left without any admixture of modern evil. It ought to be a kingdom all of its own. In Fowey all is courtesy and good reason for the chance sailing man . . . and I have never sailed into Fowey or out of Fowey without good luck attending me.'*

HILAIRE BELLOC obviously rather liked Fowey, but he was not alone among literary men and women to succumb to the romantic charm of this delightful town and deep narrow river. Daphne du Maurier lived here for many years, Sir Arthur Quiller-Couch, 'Q' immortalised it in his famous sagas as 'Troy Town', and his friend Kenneth Grahame also used it in 'The Wind in the Willows'.

Approaches

The entrance, although only a cable wide at its narrowest, is deep and easy except in strong onshore winds and ebb tide. Keep to the middle of the channel, leaving the white cross on Punch Cross rock and the triangular white beacon on Lamp Rock to starboard, and high St Catherine's Point, with its small fortress, and narrow Readymoney Cove to port, where anchoring is prohibited because of underwater cables. At night, the inner lighthouse on Whitehouse point, midway up the western shore, gives a sectored light (Iso RWG 3s) and the white sector leads straight in. Just upstream, the only other navigation lights are (2FR vert) lights at the end of the Polruan ferry landing which should be left to port and (2FR vert) light on Fowey Town Quay. There are (2FG vert) lights on Polruan Quay. The only possible problems, day or night, are likely to be a large ship emerging, or fluky winds. If departing at night there are two useful clearing lights for the entrance that only show into the harbour – St Catherine's Point (FR) visible 150° – 295°, and Lamp Rock (Fl G 5s) visible 010° – 205°.

Moorings

Once past Polruan fort, the river widens, and the moorings, then the village of Polruan itself, will come into view along

Fowey harbour entrance is very elusive from the east – the isolated white house is a good indication that you are almost there. Below: from the south, Fowey is much easier to find

the eastern side of the harbour, with the mouth of Pont Pill, a wide creek at their upper end. The main river runs northwards, with Fowey stretching along its western side in neat terraces climbing the steep hillside overlooking the harbour. Whitehouse Quay is a small pier on the Fowey shore where a small passenger ferry runs across to Polruan, and a little further upstream the low black and white twin gabled building with a veranda and large flagstaff is the Royal Fowey Yacht Club.

There is an average of 7m at LWS dredged right through the centre of the harbour, and plenty of water to within half a cable of the shore as far as Town Quay, which dries at LW, and is used by local fishing and trip boats, easily located by the tower of St Fimbarrus Church behind it, and the prominent King of Prussia hotel. Albert Quay, a short distance beyond, has a long **short stay visitors pontoon** with a bridge to the shore, which is dredged and accessible at any state of the tide, although further inshore it dries at LW. You may berth alongside this pontoon free of charge for up to two hours, which is very handy for shopping, landing crew or taking on fresh water with the hose provided. *Be warned, though, boats abusing this excellent facility, using it for an overnight stop, for instance, will be charged five times the normal mooring rate in the harbour.*

Fowey waterfront. Albert Quay, Harbour Office and short stay pontoon on extreme right

Opposite, there are usually a number of large ships moored fore and aft in the centre of the stream, waiting to berth at the china clay wharves upriver, and although this narrow waterway seems at first appearance an improbable place for much commercial activity, you have now entered the ninth largest exporting port in England, shipping over 1.5 million tons of china clay a year, and handling nearly 850 ships. Most average between 7 and 8,000 tonnes, but the largest to date, the Finnish vessels *Astrea* and *Pollux*, were nearly 17,000 tons, and 540 ft in length. It is fascinating to watch the apparent ease with which these vessels are manoeuvred to their berths, in particular the technique of 'drudging' them stern first up the river, towing from aft, with their anchors dragging on a short scope to hold the bows in position.

Considering that over 6,500 yachts also visit the port annually, somehow the commendable Harbour Authority manages to achieve a remarkable balance between run-

ning their intensive commercial activity, and catering for the large numbers of pleasure craft that pour in during the summer. Facilities have been improved considerably in recent years, a continuing policy, and the main area for visitors to head for is the 'swinging ground' just opposite the Royal Fowey YC. You can be forgiven if you think you have arrived on the 'Magic Roundabout' – during the season the Harbour launch, *Zebedee*, and the Harbour Master's dory *Dougal* are on the water seven days a week from 0830 until dark, and in the afternoon they keep a lookout to assist new arrivals. Although the Harbour Office and launch both monitor VHF channel 16, and should be called on their working channel 12, either *Fowey Harbour Radio* or *Fowey Harbour Patrol*, the landlocked nature of the harbour tends to make reception poor from outside, and you will probably not make contact until you are in the entrance when you will have been spotted anyway. The Harbour Office has a direct land line VHF link to Brixham Coastguard.

There are 18 swinging visitors' moorings for up to 16m LOA along the east side of the river just upstream of Pont Pill and a 36m visitors' pontoon. Another identical 36m pontoon is in the mouth of Pont Pill, and within the creek, which is dredged to 2m for some distance, there is another 18m pontoon astern of the fuel barge. There are also fore and aft moorings available here, very handy if you want a bit peace away from the mainstream and ideal for leaving a boat if stuck for time or weather. All the visitors' moorings are either yellow or white, clearly marked 'FHC Visitor' and there is also a floating rubbish skip close by which should be used in preference to taking it ashore.

A short distance beyond Albert Quay pontoon there is another short stay pontoon (max 2 hours) off Berrill's Yard which is being developed by the Harbour Commissioners, with new toilets/showers and a laundry planned for 1995, along with rubbish and marine toilet disposal facilities. Finally, half a mile upriver, there is more pontoon berthing for visitors in the entrance to Mixtow Pill which is particularly useful in bad weather. Note, however, that the inner part of this pontoon is private and normally reserved for local boatowners.

Anchoring is not allowed in the fairway, near underwater cables or close to landing places, and only possible with the Harbour Master's permission to the east of the swing-

There are plenty of pontoons for visitors but you will still have to raft in season

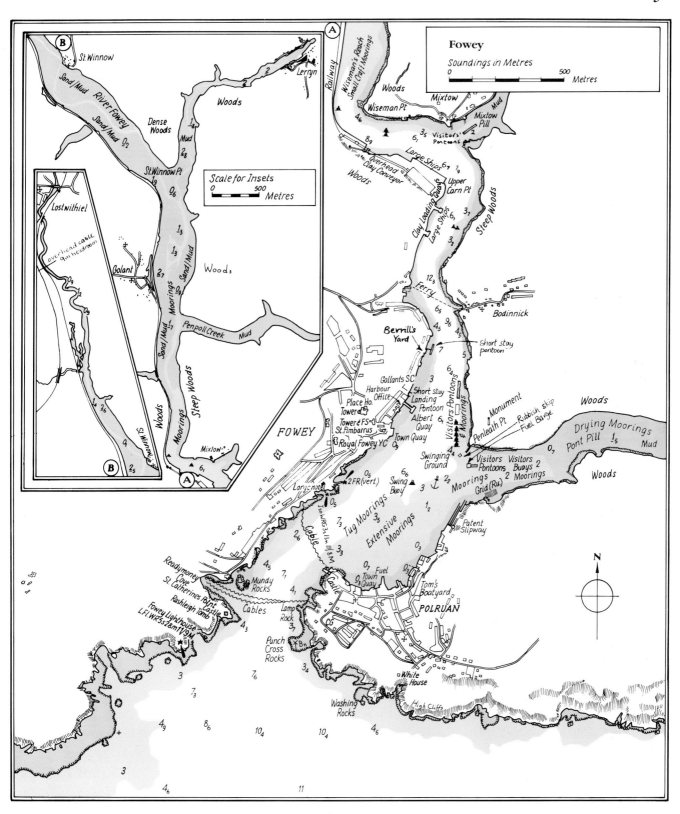

ing ground, off the entrance to Pont Pill, but this sometimes entails having to move when larger vessels are being manoeuvred. As in the River Yealm, the difference between dues for anchoring, and dues for lying to an FHC facility, only amounts to £1 for a 30 ft boat, which is an obvious incentive to use them, and well worth it for the extra peace of mind.

Facilities

Landing is easy at the Albert Quay pontoon where dinghies can be left on the inside and also at the Royal Fowey YC, pulling them clear of the steps, remembering that the foreshore here dries at LW. Members of other clubs are welcome to use their facilities, where there are excellent showers, meals by arrangement, and a good bar. Here, among the various burgees, is one that, for me, a dyed in the wool romantic, always induces a certain nostalgic twinge, as it belonged to *Moonraker of Fowey* and was presented to their club by Peter and Anne Pye.

The Fowey Gallants Sailing Club just upstream from

83

Rame Head to The Manacles

Fowey – 'the harbour of harbours'. Royal Fowey Yacht Club on the right

Albert Quay also extends a friendly welcome to visiting yachts, although facilities here are limited to a bar, with food at weekends.

The alternative to your own dinghy is the popular water taxi service which runs from Albert Quay pontoon, and can be hailed on VHF channel 06, call sign *Fowey Water Taxi*. Operating seven days a week in season from about 0800 until the pubs close, for a mere £1 a head this does away with the problem of worrying about the dinghy, and with a sheltered cuddy, it's particularly enticing when it is pouring with rain and blowing.

The town is essentially just one main shopping street, very narrow in places, with houses rising up from it and still very much as Sea Rat described it in 'The Wind in the Willows' :

'. . . the little, grey sea town I know so well, that clings along one steep side of the harbour. There, through dark doorways you look down flights of stone steps, overhung by great pink tufts of valerian and ending in a patch of sparkling blue water. The little boats that lie tethered to the rings and stanchions of the old seawall are gaily painted as those I clambered in and out of in my own childhood; the salmon leap on the flood tide, schools of mackerel flash and play past the quay-sides and foreshores and by the windows the great vessels glide, night and day, up to their moorings or forth to the sea. There, sooner or later the ships of all seafaring nations arrive; and there, at its destined hour the ship of my choice will let go its anchor . . .'

Not quite as grey today, as most of the buildings are brightly painted and very well preserved, it is also undoubtedly busier than when Kenneth Grahame was writing, with cars and people squeezing through the narrow bottlenecks. Nevertheless, it is still a delightful place, and spared the over-commercialism of so many other Cornish harbours, is perhaps a reflection on the fact that such a large proportion of its visitors come in by sea, for the atmosphere certainly reflects this.

Belloc's eulogy continues, '. . . *whatever you may need in gear is to be had at once'* – which is certainly true except perhaps on Sundays, with everything that a cruising crew might require from basics like food, drink, launderette and banks, to sailmakers, boatyards and chandler, and a Spar Foodstore open 0800-2100 daily, 0900-2100 Sundays. Diesel can be obtained alongside the fuel barge 0900-1800 daily in season but petrol remains a problem, and can only be obtained from a garage a good ½ hour walk from town. The Harbour Staff can usually help with small quantities and they can also arrange for larger amounts can be delivered in cans by the garage.

There are a variety of options for eating ashore from cafes and fish and chips to far more up-market bistros and restaurants, and several good pubs with food.

Polruan

Polruan, clinging to an even steeper hillside than Fowey, can provide all the basics. It is an attractive, far quieter little village, where, sensibly, visitors' cars are prohibited in season, and left parked at the top of the hill. Groceries are available at the 'Cottage Stores' seven days a week, there is a butcher, another chandler, a couple of pubs, and a waterside bistro/ bar. C Toms & Sons Boatyard is still very much involved with traditional boatbuilding, and often a peep into their large shed reveals a big wooden fishing boat under construction. Their slipway, one of the largest in the area, can take vessels up to 28m, and 100 tons and they have a 30 ton crane. The Harbour Authority also has a large slipway at Brazen Island further along the Polruan shore in their own repair yard which is used to maintain their tugs, dredger and barges, but private vessels up to 450 tonnes and 30ft beam can be hauled by arrangement. It is also possible for smaller craft to dry out alongside the wall here for repairs with the Harbour Master's permission.

The major sailing event of the year, **Fowey Regatta Week**, takes place during the third week in August, when the harbour is packed to absolute capacity, with feeder races from Plymouth and Falmouth, including a large contingent of their gaff-rigged working boats. Racing takes place every day both outside and inside the harbour, including the very close competition among the local one design class, the *Troy's* – a colourful fleet of 18ft three-quarter decked bermudian sloops, which sport a distinctive 4 ft bowsprit; first built in 1929, 18 survive today. The spectacular finale of the week is the Harbour Race for the Falmouth Working Boats. However, if you are seeking peace and quiet, be warned that this is the one week to avoid Fowey altogether.

Historically, Fowey has a wild and romantic past, which started when the port began to develop in the 12th century after the previous harbour at Lostwithiel, six miles inland, began to silt up. By 1346 it was able to supply 47 ships and nearly 800 men for the siege of Calais, more than any other port in England, and it seemed to give the Fowey men a particular taste for adventure. Little more than pirates, they continued to wage their own private war against the French long after hostilities had officially ceased; daring raids across the Channel and bloodthirsty skirmishes not only earned them a lot of plunder but also the nickname of 'Fowey Gallants'.

The French hit back in 1457, raiding and setting fire to the town, forcing its inhabitants to take refuge in Place House, the seat of the Treffry family, the prominent large house with tall towers just behind the Church. However, this attack only stimulated the activities of the Gallants, one of whom, John Wilcock, seized no less than 15 French ships in as many days, becoming a source of great embarrassment to King Edward IV, who sent a message to Fowey, 'I am at peace with my brother of France'. Not impressed the Gallants cut off the unfortunate messenger's ears and nose, an act of defiance that resulted in considerable punishment for the town. The ringleaders were hanged, goods seized, ships distributed to other ports and the huge protective chain slung from the forts at the harbour mouth was removed. Duly chastened, the wild men of Fowey returned to more peaceful pursuits, fishing, shipbuilding, trading and, of course, smuggling . . .

The upper reaches

Deeper draught boats normally remain in the lower part of the harbour except, perhaps, when it begins to blow from the south-west when, in spite of its landlocked nature, conditions can sometimes become very uncomfortable. Perfect in so many other ways, this has to be the one major drawback to Fowey, for a very annoying swell sets in off Polruan, and with a gale of wind against the tide in the harbour it can get surprisingly rough and very rolly at times.

Traditionally, boats used to run upriver and anchor in **Wiseman's Pool**, but it is now so full of moorings that this is no longer possible. Sometimes a few are available, so it is worth contacting the Harbourmaster for his advice but the best option is now **Mixtow Pill**. This is dredged to 2m LAT and there is a 30m pontoon available for visitors, with space for at least six 30 foot boats, and excellent shelter in all weathers. The inner pontoons are reserved for local boats and must not be used, except when directed by the Harbour staff. The foreshore on either side of the creek is private and landing is not allowed. Owners of bilge keelers can, of course, always sneak off to dry out comfortably right out of the way in the upper reaches!

Heading upstream, the river narrows above the town, steep and wooded on the eastern bank past the pretty hamlet of Bodinnick, and 'Ferryside', the prominent house by the water where the du Maurier family were brought up, but keep a lookout for the ferry itself which carries cars across to Fowey. Beyond, the extensive clay loading wharves appear along the west bank, always a fascinating stretch of river with ships of many nationalities berthed alongside, where everything, including the surrounding trees is wreathed in fine white dust, like a gentle fall of snow. The deep channel turns sharply to port at Upper Carn Point opposite the entrance to Mixtow Pill, and particular care should be taken to look out for large ships rounding this bend. At the end of this short reach, overlooked by high woods, the docks finish, and the channel turns back to the north round Wiseman Point, a wooded rocky promontory, and, thick with moorings, Wiseman's Pool appears.

Although the river is still navigable for small craft on a good tide as far as the small country town of **Lostwithiel** (but limited by an overhead power cable 1 mile downstream with 9m clearance), beyond Wiseman's Reach it dries almost completely, a mixture of sand and mudbanks. Shoal draught boats and dinghies can explore on the flood, and bilge keelers can anchor and dry out off **Golant**, where there are a large number of drying moorings and a pleasant little village with a general store, Post Office and a pub, the Fisherman's Arms. There is an FHC office here, and a public telephone by the quay, and also a handy walkway along the embankment as far downriver as Wiseman's Pool. Beyond Golant, Lerryn Creek bears away to the north-east, steep-to and wooded, this and the main channel to Lostwithiel are both marked by R and G poles. There is a 'Spar' grocer, with off-licence, and the 'Ship Inn' at **Lerryn**, with toilets and a phone by the head of the creek. At St Winnow, on the way to Lostwithiel there is a fine church in a lovely setting close by the waters' edge and, nearby, a fascinating museum of farming.

If the weather does turn and you are stuck in Fowey for a few days, apart from the museum and aquarium, the real thing to take advantage of is the number of fine walks in the area, particularly the Hall Walk which skirts a large part of the harbour. Ideally, leave the dinghy in Fowey, catch the ferry to Bodinnick where the walk is signposted on the right halfway up the steep hill leading out of the village. Far from arduous, the grassy track gently follows the contour line right round Penleath Point to the large stone memorial to Sir Arthur Quiller-Couch, where there are magnificent views of the harbour, then on through the woods and wild flowers above Pont Pill, and down to the hamlet of Pont with its old quay and water mill. From here the track climbs another wooded hillside with more lovely views all the way back to Polruan where you can catch the ferry back to Fowey and a cream tea, perhaps. . .

Another favourite is the walk to seaward of Fowey along the road to Readymoney Cove where a track leads up on to St Catherine's Head, pausing, to look at the old fortress, built by Henry VIII in 1540, with fine views into the harbour. Above it, the curious structure that can just be seen when entering from sea, two granite arches surmounted by a Maltese cross, looking like the top of a huge crown, is, in fact, the tomb of William Rashleigh, his wife and daughter, descendants of Charles Rashleigh who not only built the port of Charlestown, but also 'Menabilly' on Gribbin Head. For many years this was the home of Daphne du Maurier, and supposedly the house she immortalised in her novel *Rebecca*. Following the coast path out towards Gribbin Head you skirt the private grounds and woods surrounding Menabilly where there are spectacular views of the harbour mouth, coast, and particularly lovely scenery around Polridmouth Cove just before you reach the huge beckoning daymark, silent witness to so many departures and arrivals, such as *Moonraker's* at the end of her voyage to the Pacific, forty days out from Bermuda . . .

The sun soaked up the haze, and the town of Fowey was just out of sight. In an hour or two, or three or four, a breeze would come and we should sail in through the Heads into the harbour from which we had set out three years ago. What changes should we find, I wondered, and how should we take to living on the land? Would Christopher pad about the city in barefeet and bowler hat, and should we be content with creeks?

My thoughts were disturbed by the sound of a vessel's engine and a boat came up that was familiar. She stopped, her sails casting their shadows upon the water. Her people welcomed us.

'Come aboard,' I said, 'and have some coffee.'

And I hurried down to start the Primus.

Presently Anne looked out.

'Hullo,' she said, 'where have they gone to?'

'They wanted to get in said Christopher. 'They've already been two nights at sea.'

PORT GUIDE: FOWEY
LOCAL TELEPHONE CODE: 01726

Harbour Master: Captain Mike Sutherland, Harbourmaster's Office, Albert Quay, Fowey (Tel: 832471 /2472). Harbour patrol in season
VHF: Channel 16, working 12, call signs *Fowey Harbour Radio* or *Fowey Harbour Patrol* 0830-2100 in season
Mail drop: Harbour Office or RFYC
Emergency services: Lifeboat at Fowey. Brixham Coastguard

Anchorage: Off entrance to Pont Pill, clear of moorings and coaster swinging ground, only with Harbourmaster's permission
Moorings/ berthing: Harbour Commission – short stay pontoons (up to two hours) off Albert Quay and Berrill's yard.18 deep water moorings, two visitors' pontoons near Pont Pill and additional pontoon and moorings within Pont Pill. Pontoon in Mixtow Pill
Charges: For 30 ft boat per night inc VAT, at anchor, £5.50. On FHC facility, £6.50
Dinghy landings: Albert Quay pontoon. RFYC. Inside quay at Polruan
Water taxi: From Albert Quay pontoon, Easter-Sept, daily 0800 until pub closing time, VHF channel 06, call sign *Fowey water taxi* or hail
Marina: None
Phones: Near Town Quay, RFYC, Fowey Gallants, outside Post Office
Doctor: (Tel: 832541)
Hospital: (Tel: 832241)
Churches: C of E
Local Weather Forecast: Outside Harbour Office
Fuel: Diesel from fuel barge, off Pont Pill (0900 -1800) Petrol not available alongside, see Harbour Staff
Paraffin: Outriggers, Fowey. The Winkle Picker, Polruan
Gas: Calor/Gaz, Outriggers, in Fowey. Gaz at Winkle Picker in Polruan
Water: Albert Quay and Berrill's yard short stay pontoons
Tourist information: In Post Office
Banks/cashpoints: Barclays (cashpoint). Lloyds, Midland, no cashpoints
Post Office: In main street, turn right at Albert Quay
Rubbish: Floating skip off Pont Pill, bins ashore at Berrill's yard
Showers/toilets: Berrills yard. RFYC. Public toilets on Town Quay. Marine toilet disposal, Berrills yard
Launderette: Laundry facility at Berrills yard
Provisions: All requirements including delicatessen, EC Weds or Sat. Shops usually open in season. Spar Foodstore open late and Sunday, Fowey. Cottage Stores open seven days, Polruan
Chandler: Outriggers, by Albert Quay, also Admiralty Chart Agents,(Tel: 833233) Upper Deck Marine, Albert Quay (Tel: 832287). Fowey Marine Services, Station Road (Tel: 833236). The Winkle Picker, The Quay, Polruan (Tel: 870296)
Repairs/hauling: C Toms & Sons, Polruan (Tel: 870232). Fowey Boatyard, Passage Street, Fowey (Tel: 832194). W C Hunkin & Son, Bodinnick (Tel: 832874)
Drying out: By arrangement with Harbour Master
Marine Engineers: C Toms & Sons (Tel: 870232)
Electronic engineers: Fowey Marine Electronics, Station Road (Tel: 833101). Skywave Marine,St Austell (Tel: 70220)
Sailmaker/repairs: Mitchell Sails, North Street, Fowey (Tel: 833731)
Transport: Buses to main line railway at Par
Yacht clubs: Royal Fowey Yacht Club, Whitford Yard, Fowey PL23 1BH (Tel: 832245). Fowey Gallants Sailing Club, Amity Court, Fowey (Tel: 832335)
Eating out: Very good selection for size of town. Fish and chips to bistros/restaurants
Things to do: Fowey Museum, Aquarium, both in middle of town. Good walking, Fowey and Polruan
Special events: Fowey Regatta/Carnival, third week in August

MEVAGISSEY AND ST AUSTELL BAYS

Charlestown: HW Dover –0555.
Charts: BA 31. Stanford 13. Imray C6
Hazards: Inner floating basin entered by narrow entrance and tidal lock, and narrow entrance. Dangerous in onshore winds. Drying rocks to south and east
Mevagissey: HW Dover –0600
Charts: BA 147, 148. Stanford 13. Imray C6
Hazards: Inner harbour dries. Black rock off North Quay(unlit). Busy fishing harbour. Dangerous to approach in onshore wind, and outer harbour very exposed in east gales
Gorran Haven: HW Dover –0600
Charts: BA 148. Stanford 13. Imray C6
Hazards: Shallows extend well to seaward. Cadycrowse rock east of pier (unlit). Fine weather anchorage only and dangerous in onshore wind

Mevagissey and St Austell Bays are well sheltered from the west with depths reducing gradually and weak tidal streams. Tywardreath Bay is the area in the extreme north of St Austell Bay, and **Polkerris**, a small fishing cove on the eastern side, has a good anchorage off it in easterly winds. To reach it from Fowey, there are no hazards more than two cables from the shore, once the south cardinal 'Cannis Rock' buoy has been passed, south-east of Gribbin Head, and you can safely follow the coast this distance offshore around the western side of the Gribbin for a couple of miles until the houses and small harbour wall come into sight. Anchor about three cables to the south-west of the pier where there is about 1.5m at LW. The whole bay dries out completely, but boats able to dry out can anchor inside the harbour or lie alongside the pier. It is an unspoilt little hamlet tucked snugly beneath the cliffs and trees with a high wall along the sandy foreshore protecting it from the seas which roll in unchecked in southerly gales. Once a busy fishing harbour, but now little more than a cluster of houses, there is a pub, and a cafe in the old lifeboat house which was closed down in 1922 when the station was transferred to Fowey.

Par, with its four conspicuous chimneys, further to the north-west, is approached over a large area of drying sands. Tidal and privately owned the harbour is a busy china clay port swathed in white dust, of no interest to visitors and entry is only permitted in an absolute emergency. Midway across the bay on the outer edge of Par Sands is Killyvarder Rock, which dries 2.4m LAT and is marked by a red beacon, south of which there is often quite a cluster of coasters at anchor in the clay-tinged turquoise water, waiting to berth at Par.

Charlestown, a former china clay port two miles south-west of Par is, however, of much more interest. Privately owned since 1993 by Squaresail Shipyard Ltd the inner floating basin is used as a home base for its fleet of square rigged sailing ships, like the *Kaskelot*, and is entered through a very narrow entrance and tidal lock, yet

Visitors must lie alongside the South Pier at Mevagissey

another masterful bit of engineering by John Smeaton, who built the third Eddystone lighthouse. It has become a popular port of call for other traditional and classic sailing vessels and the sight of them lying here is an evocative step back into the past. Not surprisingly, this unspoiled 19th century harbour was used by the BBC for episodes of the *Onedin Line* and several other companies for location filming. A few pleasure craft are based here, and it is possible to lock into the basin on the tide by arrangement with the Harbourmaster 24 hours in advance (Tel: 01726 70241/67526). Alternatively, in offshore winds anchor just east of the piers and dinghy ashore for a visit. The harbour mouth dries completely, well beyond the outer breakwaters at Springs; otherwise land inside at the slip to the left of the lock gate. Apart from the harbour interest, the village of Charlestown has a large Shipwreck and Heritage Centre with one of the largest collections of shipwreck artifacts in the country, an audio visual display and life size animated scenes guaranteed to keep the children amused. There is a cafe in the centre, with Post Office/store, pubs and restaurants all close by.

St Austell Bay has many fine sandy coves and beaches,

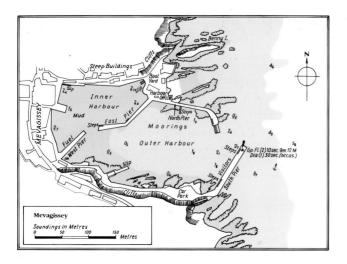

Mevagissey
Soundings in Metres
0   50   100   150
                         Metres

and with no dangers extending more than a cable from the shore, in favourable weather it is possible to sound in and anchor off most of them. Robin's Rock, just south of Porthpean dries, and there is an isolated rock drying 0.1m LAT ½ cable off Ropehaven beach.

Black Head, a bold rocky point with a grassy summit 46m high, forms the division between St Austell and Mevagissey Bays, and the fishing port of Mevagissey is two miles further to the south-west. Midway between them is the lost harbour of **Pentewan**, at the northern end of the long stretch of Pentewan sands, another tidal basin once approached by an artificial channel that has now completely silted up.

Mevagissey, or 'Mevva' as it is known locally is the best known harbour in the bay, a classic Cornish fishing town with good shelter in the prevailing winds, but exposed in easterlies, when it can be dangerous to enter and very uncomfortable inside. It comprises an outer harbour, where fishing boat moorings take up most of the available space, and an inner harbour which dries completely and is restricted to local boats only. It is predictably popular with holidaymakers, and becomes very busy in season, but in spite of a certain amount of commercialism it retains much of its character; pastel cottages, slate hung, with cement washed roofs creep up narrow alleyways to overlook the harbour, rather like a small version of Brixham. It is well worth a visit providing there is no inkling of the wind turning to the east but remember, however, that fishing is still a very important part of the town's life, with quite a large trawling and potting fleet, and also a number of trip and angling boats. Commercial activity takes precedence at all times but visiting boats are very welcome as long as they are prepared to work around the port activity, and do not impede the fish landing and other boat movements.

The approach is easy, with no off-lying dangers except the Gwineas rocks if arriving from the south. Steer for the white lighthouse (Fl (2)10s 12M) on the South Pier, keeping close to it, as a shallow sandy bank, drying 0.3m LAT extends east from the end of the North Pier which also has a 6 foot wide concrete apron around its base which covers at half tide, marked on the southern corner by a pole with conical green topmark. The outer harbour is available even at LWS, with depths alongside the inside of the outer end

of the South Pier to just over 2m. Visitors normally berth here, rafting alongside fishing boats between the steps. Boats able to dry out can sometimes lie on 'Sandy Beach' on the seaward side of the inner West Quay with the Harbourmaster's permission. *No visiting boats are allowed in the inner harbour.*

Captain Hugh Bowles, the Harbour Master listens on VHF channel 16 and works on channel 14, 0900-2100 (April-Oct). His office (Tel: 0726 843305) is a clearly marked white building on the north pier. Do not leave your boat completely unattended in case you have to move. Because of the restricted space anchoring is not allowed, but there is sometimes a possibility of a mooring being available in the outer harbour. Wherever you lie, charges for a 30 ft boat are £5.00 per night, inclusive of VAT.

Renowned for smuggling activity and one of the largest centres on the south coast of Cornwall for the 18th and 19th century pilchard fishery, Mevagissey was a rough, unruly place in its heyday, reeking with the stench of fish, with huge landings, sometimes in excess of 30 or 40 thousand fish per boat, all of which were counted by hand.

There is an excellent small museum close to the Harbour Office, a fine old wooden building dating from 1745 that was once part of a boatyard, and right next to it, another 'real' boatyard remains where John Moore still builds beautiful traditional wooden fishing boats.

Although busy in summer, a wander around the town is pleasant, and most requirements can be found, including a Lloyds bank open 0930-1230 during the season, and even an aquarium. There are plenty of cafes, fish and chips and a few restaurants, and some good pubs. A small amount of chandlery is available, Calor and Camping Gaz, and diesel from the fuel berth at the inner end of the South Pier if you contact the Harbourmaster. Water is only available in cans from a tap by the Harbour Office or the jetty in the inner harbour. Petrol can be obtained in cans from a nearby garage, and there is also a marine engineer if needed.

South of Mevagissey, **Gorran Haven** is an attractive fine weather anchorage, sheltered in westerly winds, with a drying small boat harbour and a gently shelving clean sandy beach. Approaching from the north keep midway between Gwineas Rocks and the low promontory of Chapel Point, sound in and anchor two cables east of the sea wall; in settled weather bilge keelers can dry out comfortably closer inshore. Just a narrow street of old fishermen's cottages, there is a Post Office, general store (open Sundays), several cafes, a pub, and the start of a lovely coast path up to Dodman Point.

FALMOUTH

Tides: HW Dover +0600. Range: MHWS 5.3m – MHWN 4.2m. MLWN 1.9m – MLWS 0.6m. HW Truro approx 8 mins after HW Falmouth. Streams attain over 2kts in upper reaches at springs

Charts: BA: 32,154,18. Stanford: 13. Imray: Y58

Waypoint: Eastern dock breakwater end 50°09′31N 05°0290W. Black Rock E. Cardinal buoy 50°08′′66N. 05°01′77W

Hazards: Black Rock (unlit, but outside lit buoyed channel). St Mawes Buoy/Lugo Rock (unlit). Governor Buoy (unlit). Busy commercial port, beware shipping movements in vicinity of Docks, Penryn River, and upper reaches of Fal. Large areas of upper reaches dry. Beware pot and net buoys

'*almouth for orders'* – just a few words that evoke so much, and although lofty spars, square yards and billowing spreads of canvas no longer grace the Western Approaches and deep laden rust streaked hulls waiting on the whims of commerce have long ceased to swing at anchor in Carrick Roads, Falmouth, by virtue of its far westerly position, is still a major port of landfall and departure for many of the smaller sailing vessels that now cross the lonely oceans of the world and the ensigns of many nations are a familiar and romantic sight around the harbour during the summer months.

For the majority with less exotic cruising ambitions, the River Fal and its tributaries offer everything that a yachtsman might require. As my own adopted home port, in spite of an inevitable degree of bias, most visitors would probably agree with me that it has to be one of the finest sailing areas in England, and certainly, it ranks as one of the great natural harbours of the world.

Approaches

Flanked by St Anthony Head to the east, grass topped with low cliffs and a white lighthouse (Occ WR 15s 22M, Red Sector 004° – 022°) built in 1835, and Pendennis Point and castle nearly a mile to the west, the entrance is easy and can sensibly be made in any weather, although a gentle ride is hardly likely to be encountered in a southerly gale with an ebb tide. There is, however, no such thing as total perfection, and although deep, the entrance does have one

notable hazard, Black Rock, perversely right in the middle, which uncovers at half tide, and is marked by a distinctive but unlit black conical beacon with an isolated danger mark (pole and two spheres) on top. You can enter either side of Black Rock, the buoyed deep water channel for commercial shipping to the east, the western channel, quite safe for small craft has a least depth of 6m; do not, however, be tempted to pass too close to Black Rock as shallows extend nearly 200m north and south. Strangers arriving at night should keep to the deepwater channel which is well lit, with 'Black Rock' east cardinal buoy, BYB (Q (3) 10s) to port, and 'Castle' conical green (Fl G 10s) to starboard. From here, steer up towards 'West Narrows', red can (Fl (2) R 10s), but midway between them you can bear away towards the distant eastern breakwater end (Fl R 2s). This will ensure that you safely pass north of the unlit 'Governor', BYB east cardinal buoy.

Pendennis Castle, initially completed in 1543, was part of Henry VIII's massive chain of coastal defences from Milford Haven to Hull, built in response to fears of religious inspired invasion from Europe. His daughter Elizabeth instigated the building of the large outer walls during the war with Spain, after Penzance, Newlyn and Mousehole were raided and burnt in 1595. But the awaited attack never came, and Pendennis never fired its guns in anger. Preserved today by English Heritage, this, and its counterpart at St Mawes are open daily during the summer, and the walk up to the headland along Castle Drive, with a

Falmouth entrance from the south-east. St Anthony light, Pendennis Point and castle, with Black Rock in centre

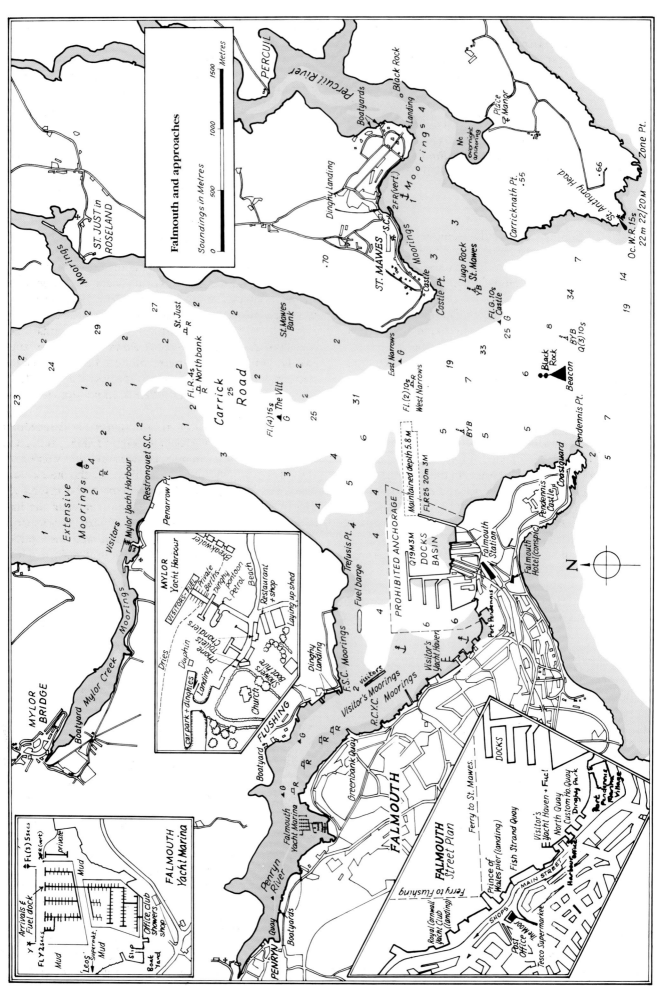

grandstand view of the docks, estuary, and Falmouth Bay is definitely recommended.

Just below the castle ramparts, the squat modern building on the hillside is the Falmouth Coastguard Maritime Rescue Co-ordination Centre, now regularly involved in rescues on the other side of the world thanks to satellite technology. Monitoring channel 16, the local working channels are 67, 10, 73, call sign *Falmouth Coastguard*. It should not be confused with *Pendennis Radio*, the BT VHF coastal radio station, which should be called direct for link calls on working channel 62.

Moorings

With anchorages, moorings, berths alongside and a marina, the Fal has a wide if somewhat bewildering variety of options for visiting boats, but Falmouth itself does tend to be the first stop, and with ample deep water well into the inner harbour there are no problems entering. Follow the line of northern arm of the docks, keeping a wary eye on shipping movements in and around them, particularly when rounding the western end which should be given a wide berth – tugs, ferries, fishing boats, and sizeable ships have a nasty habit of emerging, often at speed.

Dating from around 1860, Falmouth Docks can dry dock ships up to 100,000 tons for repairs and they are also an important bunkering facility. Within the Docks 'Pendennis Shipyard' is an independent company specialising in the building and refit of large luxury yachts, which can often been seen lying in their wet basin, on the eastern side of the Docks. They have a 500 ton slip and 80 ton hoist and are internationally renowned for their high quality work.

Opening ahead and to port the picturesque town of Falmouth spreads along the hillside overlooking the mass of moorings in the inner harbour. Bear round past the western side of the docks, and steer for Custom House Quay dead ahead, where you will find the main anchorage and the Falmouth Harbour Commissioners' ever-popular **Visitors' Yacht Haven**. Here, May – September, there is 172m of pontoon berthing for 40 boats, up to a maximum 12m LOA, 1.8m draught, and at the height of the season you will invariably have to raft. *The pontoon area is dredged and the water to the north and west of it very shallow at MLWS when it should be approached with caution. Hold your course for Custom House Quay until the pontoon arms are well open abeam, then make your final approach directly from the east.* Charges for a 30 footer are £8.40 per night inc VAT. The Harbour Office, call sign *Falmouth Harbour Radio*, monitors VHF channel 16, and works on 12, as does their patrol launch *Killigrew* which is usually on hand to assist during working hours, when there is also a Supervisor on the Yacht Haven. There are water hoses on the pontoons and adjoining it 'The Boathouse' fuel barge has diesel, petrol and water for customers, (open daily 0900-1800). An alternative for diesel, particularly for larger vessels is the *Falmouth Industry*, a long blue barge moored just south of Trefusis Point on the far side of the harbour. This is open Mon-Fri 0800-1630, Sat 0800-1200, call sign *Falmouth Industry* VHF channel 16, otherwise berth alongside.

There is a rubbish skip on North Quay, and just around the corner, the Yacht Haven Amenity Building has good showers, toilets and laundry facilities, in what was previously the local RNLI shore base. This, and the lifeboat, are now located ajdacent to **Port Pendennis**, the marina village development in the south-eastern corner of the harbour. Though mostly reserved for local boats visitors' pontoon berths are sometimes available here and it is a particularly safe spot if you have to leave your boat for a while.

Falmouth, looking to Flushing from Custom House Quay. Visitors' Yacht Haven in centre, anchorage on the right

Rame Head to The Manacles

Access is through an automatic dock gate approximately 3 hours either side of HW and is controlled by a traffic light. Only enter when the green light is showing; the gate closes as soon as it turns to red. If closed, there is a large waiting pontoon in the approach to the gate. Call *Port Pendennis* VHF channel 80 or 37, or (Tel: 211819)

The Yacht Haven is by far the most convenient alongside berth in town as it is not far from the main shops, restaurants and bistros. 'The Bosun's Locker/Marine Instruments', a large chandler and Admiralty Chart agent is just minutes away on the waterfront; 'Millers' excellent nautical bookshop is just up the road.

Over three hundred years old, the quays surrounding the drying inner basin form an attractive focal point on the Falmouth waterfront. There are two pubs, the Chain Locker and the Quayside Inn, with tables and seats on the quay, overlooked by some fine old buildings, including the Harbour Office, and Custom House built in 1815 with its adjoining brick chimney, the 'Kings Pipe', still used for burning contraband. A summer ferry to St Mawes runs from Custom House Quay.

If you decide to anchor instead – more privacy, and less to pay, a 30 footer will cost £3.45 per night inc VAT – let go just east of the Yacht Haven, clear of the quay and local moorings, but also well away from the Docks, as anchoring is prohibited within a cable of the jetties, an area of water regularly used for swinging large vessels into their berths, when the anchorage often has to be cleared. Details of impending ship movements are displayed at the Yacht Haven Office, and the Harbour staff will give you good warning. Depths vary between 1.5m and 2.5m, and the holding ground and shelter is good in anything except northerlies, but the wash from the ferries and pleasure boats can make it fairly lively at times, and in the season it can get crowded. The best place to land is on the innermost Yacht Haven pontoon, by the linking bridge to the Quay, where dinghies can be left afloat.

An alternative anchorage will be found north-east of Prince of Wales Pier, outside the local moorings in 2-3m, but inside the fairway leading up to the Penryn River. Here too, in summer, the Harbour Commissioners have a 70m long pontoon island for visitors which can take up to 16 rafted boats to a maximum of 12m LOA, at a cost £7.68 per night inc VAT for a 30 footer.

Shelter is normally good throughout the inner harbour except in strong east, north-east or south-easterly winds when it becomes surprisingly rough as seas build across the open fetch from the St Mawes shore. Fortunately, easterly gales are rare, particularly in summer. Keep well clear of the Prince of Wales Pier as it is constantly busy with large trip boats, and ferries to Flushing and St Mawes; there is, however, a good dinghy landing in the boat harbour on the inner end of the pier which is particularly handy for the Tesco supermarket and the shopping centre of the 'Moor'.

Finally, between Prince of Wales Pier and Greenbank Quay there are 20 swinging visitors' moorings; the Harbour Commissioners, K1 to K6, and T1 to T6, with green support buoys, marked FHC, which can be booked in advance through the Harbour Office, and cost £6.92 a night inc VAT for a 30 footer. The rest, marked 'RCYC Visitor' belong to the Royal Cornwall Yacht Club and are available at £7.00 a night inc VAT. Their fine clubhouse stands on its own quay, below the elegant row of tall Regency houses along Greenbank, just upriver of the new, extensive Packet Quays waterfront development. These moorings can be reserved in advance through the Club Secretary, and visitors from other recognised clubs are made very welcome.

Established in 1871, the club is open every day except Monday and, among its many activities, organises the Azores and Back Race every four years. Facilities include showers, bar and dining room; there is a good dinghy landing and a Club launch which operates daily between 1000 and 1900 (later on Tuesday and Friday evenings when racing takes place) and visitors are welcome to use this free taxi service, call sign *Club Launch* VHF channel 37. Fresh water and scrubbing alongside the club quay is also possible by arrangement.

Packets, Quay Punts and Working Boats

On his way home from El Dorado in 1596 Sir Walter Raleigh put into the Fal and found little more than a small fishing village, known as 'Smithick', and the nearby manor house at Arwenack, home of the Killigrew family. His advice that it might make a good harbour was not instantly taken up, but by 1670, the Killigrews had got round to building Custom House Quay, which turned out to be a stroke of luck rather than foresight, ultimately transforming Falmouth into a thriving seaport of major importance. Impressed by the new quays, the Post Office decided to establish its Packet Service here in 1688. This fleet of fast, armed sailing ships, generally brigantines of around 200 tons, with appropriate names like *Speedy* and *Express*, were all privately owned and chartered to the Crown, carrying mail, bullion, and passengers, not only to Europe but as far afield as the West Indies, and the Americas, a round trip averaging 15 to 18 weeks. It was a tough, dangerous business, the valuable cargoes resulting in frequent attacks by privateers and pirates, but many owners and captains made fortunes. Around Falmouth, particularly along Greenbank and opposite, in the small village of Flushing, large elegant houses rose along the waterfront, graphic evidence of the new-found wealth; by 1800, over 40 packet ships were based in the port.

As the first port of call for most inbound shipping,

The spectacular working boats race during the summer

before the days of wireless, Falmouth often received the first news of dramatic events abroad, such as the death of Nelson. Ship's Masters were not only anxious to notify owners of their safe arrival but also to find the final destination for their cargoes – 'orders' – which resulted in much trade for Falmouth and the development of the famous 'Quay Punts' to service the deep water visitors. These deep, sturdy and mostly open boats were between 25 and 32 ft long, and ranged the approaches to the Lizard 'seeking' for ships before they arrived off the port. The particular rigours of the job produced a remarkably seaworthy craft; yawl rigged with a stumpy mainmast to enable them to sail in under the lower yards of the big sailing ships. Only one was ever lost performing her job; named *Fat Boat* she specialised in collecting the accumulated cooking fat from the incoming ships to sell ashore. Swamped off Black Head in 1904 she sank in seconds, but many of her more fortunate contemporaries were later decked in and converted to yachts which are still sailing, and a number of yachts were built along their lines. The quay punting still continues in Falmouth today with a small fleet of sturdy motor launches operated by the Falmouth Licensed Watermen.

The other traditional craft that originated in the estuary of the Fal are the Working Boats, the sailing oyster boats that are now unique in Europe, their freak survival resulting from the local bye-laws prohibiting the dredging with anything other than sail or oar. Gaff cutters varying between 20 and 32 ft in length, although mostly around 28 ft, a number of these three-quarter decked boats still work the natural oyster fishery during the winter season, drifting down tide across the banks in the upper part of Carrick Roads, and hauling their dredges by hand. During the summer, however, they race, a tradition continued by the Falmouth Working Boat Association, formed in 1979 to preserve the unique character of the class. With wooden boats, some of them well into their 80s, and several much younger, together with fibreglass hulls that have been in production since the 1970s, and even one in ferro, this impressive fleet of up to 20 boats at times, with huge gaff rigs, lofty coloured topsails and long, lethal bowsprits is one of the most spectacular sights in Falmouth. One that you cannot possibly miss, certainly if you are around at the weekend, the boats, with their large open cockpits, have room for plenty of crew, and if you feel like trying it, just ask, as someone will invariably take you along.

In marked contrast, the other distinctive Falmouth racing fleet, the Sunbeams, epitomises the elegance of the 20s and 30s, and the far off days when the mighty 'J' class raced in Falmouth Bay. They are classic three-quarter decked bermudian sloops with long overhangs, just under 27 ft overall and this immaculately maintained, colourful small fleet is one of two in the country, the other based in the Solent where the first boats were built in 1923. Racing in Falmouth is very active all year round, but **Falmouth Regatta Week,** during the second week of August, organised by The Port of Falmouth Sailing Association (POFSA) is the major event, beginning with the spectacular 'Falmouth Classics' weekend for traditional craft.

The advent of the steam ships finally removed the good fortune that Falmouth had enjoyed for over 150 years, and

in 1852 the Post Office transferred the mail service to Southampton. Wireless did away with the need to call for orders but the decline in the port was brief, for it was not long before the new docks were commenced in 1858; building ships, repairing them and cashing in on the new needs of the steamers for bunkers.

With the arrival of the railway in 1863 another new industry began to emerge – tourism. In 1865 the Falmouth Hotel was the first to rise from the fields along the seafront, and the holidaymakers had begun to arrive; today, more than ever, they are the mainstay of the town's prosperity.

Falmouth Yacht Marina. Falmouth harbour is on right

Falmouth Yacht Marina

However, the biggest boating facility in Falmouth, the Marina, is about ½ mile further upstream, on the Penryn River, beyond Greenbank Quay, which is not difficult to miss as it has 'Greenbank Hotel' in huge letters along it. From here on the river becomes shallower, but the main channel always has over 2m and is buoyed, red cans to port, and conical green buoys to starboard, the first red can being the only one lit (QR). The southern shore is particularly shallow; drying banks extend well out and are covered with oyster beds where anchoring is prohibited and you run aground at your peril.

Opposite, **Flushing** is an attractive small village with limited provisions, two pubs and a restaurant on the quay, where a small fleet of local fishing boats land their catches. The building on the lower 'New' quay is Flushing Sailing Club, which is generally only open at weekends or when evening racing is taking place; there is a bar but no other facilities. The walk from Flushing out to Trefusis Point (and right round to Mylor if you are feeling energetic) provides some fine views of the Fal and there is a pleasant sandy beach at Kiln quay, popular with locals and a regular spot for barbeques.

From downstream the Marina is hidden behind Boyars Cellars, or 'Coastlines' as it is known locally, a jetty with warehouses on the port hand where coastal petrol tankers regularly discharge, usually arriving and leaving near HW. In the channel they have complete priority and you should keep well clear. Opposite, at Little Falmouth, are the slipways and large sheds of Falmouth Boat Construction where there has been a boatyard since the early 1880s, and site of the first dry-dock in Falmouth.

Rame Head to The Manacles

The proposal to dredge out and build a marina in the muddy creek at Ponsharden was greeted with great local scepticism when it was first proposed in 1979, but the success of the 330 berth **Falmouth Yacht Marina** has long since proved all the pundits wrong, and usually there are up to 70 visitors' berths available.

The dredged approach channel provides access for boats drawing up to 1.8m at LW, although a certain amount of care is recommended at LWS. The isolated pontoon just beyond Boyars Cellars is private berthing for local boats with (2FR vert) lights on its northern end; beyond, visitors must keep to **port** of the BY pole with E Cardinal topmark (Fl (3) 5s) and proceed to the reception/fuel berth which is clearly marked on the outermost marina pontoon. Here, the berthing master will allocate a berth, or call *Falmouth Yacht Marina* ahead on VHF channel 80 or 37, particularly if you are over 13m LOA. The BY Pole with yellow X topmark (Fl 2s) beyond the fuel berth marks the northern and western edges of the dredged channel. At night this light should be kept open to port of the E Cardinal light in the final approach.

The final approach to the marina is well marked – keep the East Cardinal beacon on your starboard hand

Charges for a 30 footer are £14.50 per night inc VAT, the facilities are good, and the marina has a friendly atmosphere. There are toilets/showers, 24 hour launderette, chandler, Calor and Camping Gaz, water at the berth, and electricity can be provided if required. Diesel is available 24 hours, either call ahead on VHF, or berth on the fuel pontoon and if no one is in attendance contact the Marina Office with the phone provided. Berthing masters are on duty from 0700 – 2300 during the season, and there is also a night watchman providing 24 hour security, with security gates to the pontoons. In addition, there is a 30 ton mobile hoist and full repair facilities. Petrol can be obtained in cans from a nearby garage, and five minutes' walk away 'Leo's' large supermarket is particularly convenient for big provisioning jobs, and open on Sundays too.

The only real drawback to the Marina is its distance from the centre of Falmouth, which is about 15 minutes' walk away; there are, however, regular buses every 20 minutes and plenty of local taxis.

Penryn

Between the Marina and Penryn, ½ mile further upstream, the river dries almost completely. During the Middle Ages Penryn was a port of considerable importance until its trade was wrested away by the development of Falmouth and the gradual silting of the river. It is possible to reach Penryn quay after half flood, but it is much used by local fishing boats and the mud is sticky and deep at LW. Scenically the river does not really have a lot to recommend it, although the town is architecturally very interesting. However, if you have any probems, Penryn is where you're likely to end up as most of the main marine businesses are located in the area; if you wish to avoid going upriver by boat regular buses run from the centre of Falmouth.

On the opposite side of the creek to the marina is Port Falmouth Boatyard, which has a 120 ton slipway and l8 ton mobile hoist. Williams Boatyard, M&L Yachts Boatyard and Challenger Marine (drying pontoon berthing) are further upstream. Right at the head of the creek, a number of marine orientated businesses are located at Islington Wharf, including 'South West Sails', and it is possible to dry out alongside, although again, muddy. On Commercial Road which runs parallel to the creek you will find 'The Boathouse' chandlers, 'SKB Sails' 'Robin Curnow' outboard specialists and also liferaft repair and servicing.

St Mawes

The real beauty of the Fal is that it can provide within its limits most of the ingredients of any enjoyable cruise; magnificent scenery, safe water for sailing, and some delightful anchorages where you can escape from the crowds ashore, and find some welcome peace and quiet.

Stretching nearly three miles inland and almost a mile wide Carrick Roads is a broad stretch of water, surrounded by rounded hills, fields and low rocky shores. At the south-eastern end, opposite Falmouth, is the entrance to the Percuil River and St Mawes, which has only one hazard in the approach, Lugo rock, two cables south of the prominent castle, least depth 0.8m LAT. It is marked by 'St Mawes', an unlit YB south cardinal buoy, which should be left on your port hand when entering. The houses of the small village of St Mawes run along the northern shore and off them there are a number of local moorings, and a small boat harbour lies beyond them. There is a fairway up to the jetty end which is used constantly by the ferries and pleasure boats, so anchor well clear between it and the moorings. Land on the large slipway or beach and keep clear of the pier.

Off the main holiday track, St Mawes has remained somewhat aloof, a former fishing village with a particularly mild climate resulted in its early popularity for retirement and more recently, holiday homes, a combination that has resulted in a welcome resistance to over-development and commercialism. Built at the same time as Pendennis, St Mawes Castle is a particularly well preserved example of its type, a clover leaf formation of three immense bastions, now open to the public and surrounded by gardens with fine views across to Falmouth.

The village provides all the necessary basics: food, a Lloyds bank, Post Office and there is a small chandler, a delicatessen, cafes, restaurants and several pubs, including the popular Victory Inn. Fresh water is available from a tap

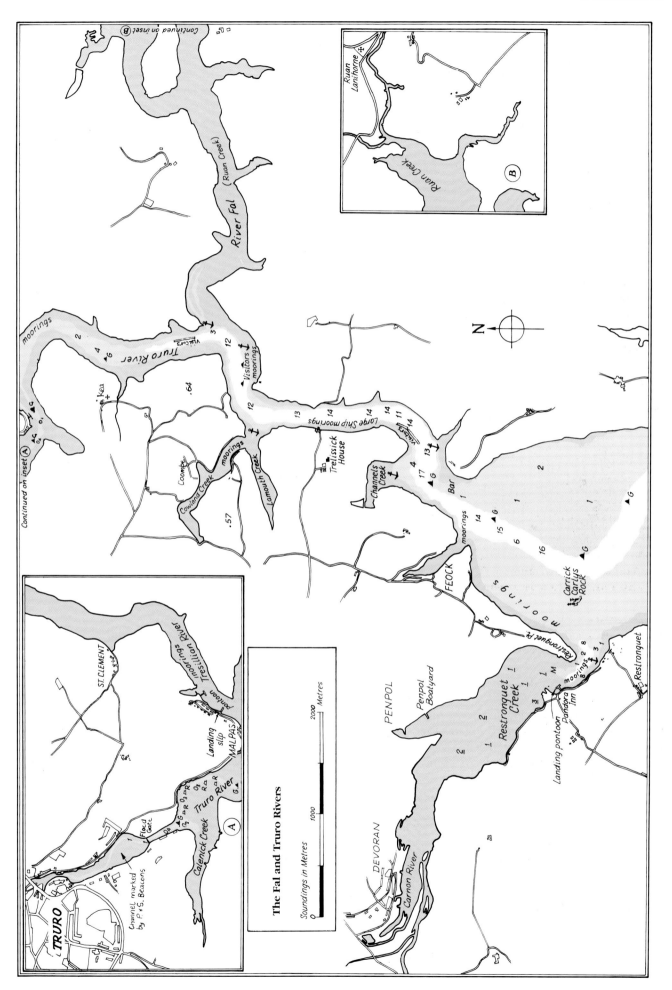

The Fal and Truro Rivers

Soundings in Metres

0   1000   2000 Metres

Rame Head to The Manacles

St Mawes from the Percuil River. Amsterdam Point on left

on the quay near Lloyd's Bank, but diesel is only available in cans from the Freshwater and Percuil boatyards further upriver. Calor Gas and some chandlery is also available from Percuil Boatyard.

St Mawes Sailing Club overlooks the quay, and visitors are welcome to use their bar and showers; they also have four red and white visitors' moorings up the Percuil off the club's dinghy store at Old Stone Works Quay, which is handy for landing, and just a short walk from the village. Ask at the clubhouse or phone ahead (Tel: 270686) for availability. Visitors' moorings can also sometimes be arranged through the St Mawes Harbour Master, whose office is on the Town Quay (Tel: 270553), or the Freshwater Boatyard (Tel: 270443).

Perfect in easterly winds, the anchorage off St Mawes is exposed to the south-west, and shelter then can only be found up the Percuil River, around the bend past Amsterdam Point, but anchoring is difficult because of the large number of local moorings. You might just find space to anchor clear of the moorings in the bight off Place House, a peaceful spot overlooked by silent woods, but the foreshore to MLWS is owned by Place and the owner does not allow boats to dry out here overnight so you will probably be asked to move. The Percuil River winds away inland, drying to a large extent, but shallow draught boats can follow it with a rising tide as far as Percuil with the local moorings all lying in the line of the deeper water. Here it is possible to anchor clear of the local moorings in mid-stream, just grounding at LWS but do not wander anywhere close inshore above Polvarth Point as oyster beds, marked by withies, lie on both sides of the river. Ideally, this is one of those places best explored by dinghy, the deserted creeks are alive with herons and other wildlife and there is excellent blackberrying late in the summer.

Mylor Yacht Harbour

Leaving the narrow Percuil, do not be lulled into a false sense of security by the broad waters of Carrick Roads for large areas are shallow at LW, as many have discovered to their surprise. However, the main channel is very deep, averaging 25m LAT, and trends towards Penarrow Point, a low promontory with a prominent pillar on the western shore, then over to the eastern bank towards St Just Creek, where there are a number of moorings. At half flood there is ample water everywhere, but upstream of St Mawes Castle, St Mawes Bank extends from the shore, least depth 1.2m LAT in places. Mylor Creek, west of Penarrow Point is a popular base for a large number of local boats which lie on extensive moorings in Mylor Pool, but the surrounding shallows in the approaches have as little as 1.2m LAT in places, and with Springs, or a draught over 1.5m, it is always best to wait until a couple of hours after LW, passing close to the red can 'North Bank' buoy off Penarrow Point. Next steer for and pass between the two smaller red can and conical green approach buoys to Mylor Pool (May-Sept). Maintain this sort of distance off the shore as you approach the moorings.

Restronguet Sailing Club, an active dinghy racing centre, is the modern building on the foreshore, and beyond it are the houses, quays and pontoons of Mylor Yacht Harbour. Formerly the Royal Navy's smallest Dockyard and base for the old wooden waller *HMS Ganges,* it was not until the late 1960s that this picturesque corner began to develop into the busy yard that it has since become.

The yard has some visitors' moorings off the pontoons, marked 'C', and these can be booked in advance, otherwise call on VHF channel 80 or 37 Mylor Yacht Harbour during working hours, or berth alongside the northern side of the outermost pontoon clear of the fuel berth and enquire at the office ashore. Approaching the pontoons beware the low floating breakwater to the east, and remember, alongside the outer pontoon there is only 1m at MLWS. The overnight charge for a 30 footer is £7.63 inc VAT for a swinging mooring or £12.92 inc VAT alongside.

Mylor Pool from the Yacht Harbour. Mylor Yacht Club is on the end of the quay

There is no public transport to Mylor Yacht Harbour but it is very self-contained, with water and diesel at the fuel berth, petrol at the end of the main quay, showers, toilets, a small launderette, and licensed cafe/restaurant. The chandler/foodstore is open seven days a week, and as well as boating clothing and an electronics engineer/supplier, the comprehensive yard facilities include a 25 ton travelift, repairs and engineer. Last but not least, the very friendly Mylor Yacht Club is on the quay and visitors are welcome to use the clubhouse and bar. The old quays of the dockyard form a sheltered and attractive setting overlooked by the lovely church of St Mylor, and some of the gravestones are a fascinating reflection of the local maritime heritage,

Mylor Yacht Harbour and moorings looking east to Carrick Roads

the peaceful resting place of Packet Commanders, sailors and oyster dredgermen.

Overhung with trees, the narrow entrance into Mylor Creek is very full of moorings but widens further in, continuing for nearly a mile to the village of Mylor Bridge, where there is a Post Office, shops and occasional buses to Falmouth. All of the upper part of the creek dries, and at Tregatreath is 'Gaffers and Luggers' boatyard, where traditional shipwright work blends harmoniously with modern technology for it is here that the gaff-rigged fibreglass working boats are produced by Martin Heard.

From Mylor Pool, it is tempting to head directly for Restronguet Creek just upstream, certain disaster if the tide is near LW as beyond the outer moorings the extensive banks almost dry at LW.

Instead, return to the main channel, heading for the moorings off St Just, a pleasant spot in an east wind, when Mylor inevitably suffers the most. There are many local moorings but a reasonable anchorage can be found just clear of them, off Messack Point or the beaches to the south of the creek. Shoal draught boats can work their way right into the creek and dry out there if space permits. Land on the shingle beach at the head of the inlet that encloses a drying tidal pool, where you will find Pascoe's Boatyard nearby. Small quantities of diesel are available here, in cans, and also Calor refills. The undoubted pride of the creek is the 13th century church. Often described as having one of the most beautiful settings in England, it nestles against the wooded hillside overlooking the peaceful creek

surrounded by a breathtaking blaze of camelias and rhododendron bushes in spring.

Restronguet

Near low water keep within the main channel, which bears north-west from St Just towards Restronguet. It is shallow between the mouth of the creek and the main channel, south of the two beacons, north and south cardinal marks, on Carrick Carlys Rock, but with a couple of hours of flood you should be able to get up to the Pandora Inn, the most likely reason for a visit to the creek. There are concentrated moorings in the deep pool at the narrow mouth of the creek off Restronguet Point, a low promontory with some very expensive looking properties along it, and the stream runs fast through the narrows. Deep draught boats can anchor east of the entrance off Weir Point before you reach the moorings. Either land on the beach and walk along the path to the pub, or take the dinghy, an easy row with the tide, and make sure you return on the ebb. Boats able to dry out, however, can go right up to the large pontoon off the pub. With its huge thatch and attractive waterside location the 'Pan' is inevitably a very popular local watering hole which takes its name from *HMS Pandora,* the ship that captured the Bounty mutineers. On his return in 1790 her commander bought this 13th century inn, and today, not only can you find food and drink at the pub, but also a very handy shower and toilet block, a small launderette, and fresh water.

The rest of Restronguet Creek widens considerably

Rame Head to The Manacles

The famous Pandora Inn is in Restronguet Creek

above the Pandora, fringed by woods and fields down to the water's edge. It is very shallow and dries extensively, but can be explored by dinghy for over a mile as far as Devoran, a small village on the northern shore where there is a Post Office and general store. A busy port in the 1880s exporting copper and tin, the alluvial mining deposits eventually caused its demise, silting the river to such an extent that it is now only accessible a couple of hours either side of HW.

The upper reaches

Beyond Restronguet, leave Carrick Carlys Rock to port, and on a falling tide keep to the deep water channel, marked by conical green starboard hand buoys. The bank to the east is particularly shallow, as little as 0.2m in places at LW, and this is the main area of the oyster fishery in winter. Opposite, there are many moorings off **Loe Beach**, shingle and sand, and popular for swimming. A few visitors' moorings are sometimes available through the cafe, otherwise anchor off. Just above the beach, Pill Creek is a narrow wooded inlet, completely taken up by local moorings.

The real upper reaches of the Fal and Truro Rivers begin at Turnaware Point, wooded, with a low shingle foreshore on the eastern bank. The change is dramatic as the wide expanse of Carrick Roads narrows into a deep waterway and the most attractive part of the river begins. Keep close to the western shore, covered in trees, and fringed with cliffs, as a notorious grounding spot, **Turnaware Bar**, extends north-west from the Point. It is clearly marked by a green conical buoy which must be left to starboard. Here the streams begin to run strongly, 2-3 knots at Springs.

Immediately north of the buoy is **Channals Creek**, a popular local anchorage, well sheltered in anything except southerlies, and out of the main tidal stream. Sound in off the edge of the deep channel to about 2m at LW. Shoal draught boats can get much closer inshore where they will ground. A fine sweep of grassland leads up from the water's edge to the impressive facade of Trelissick House, built in 1750, and now owned by the National Trust. The house is not open to the public but the gardens are, from March to October; a particular delight for lovers of hydrangeas as there are over 130 varieties. Land at the rocky point on the east side of the bay and follow the scenic footpath along the shore through the woods.

The anchorage inside of Turnaware Point is another popular spot at weekends, particularly for picnics and barbeques. There is another good anchorage nearby off diminutive **Tolcarne Creek**, which is overlooked by steep woods, just clear of the deep water. The only thing that can detract from this delightful spot is the increasing use of it by water skiers, but mid-week it is usually peaceful.

Rounding the corner there is a Truro Harbour Authority visitors' pontoon, with a rubbish bin and useful information board. Charges here are £5.00 a night inc VAT, and dues are collected by the Harbourmaster's launch which monitors VHF channels 16 and 12, call sign *Carrick Three*. Truro Harbour Office uses the same channels, call sign *Carrick One*. Beyond, the biggest surprise of the Fal comes into sight. Here, in the narrow river, surrounded by high wooded shores, you will usually find several large merchant ships laid up on fore and aft moorings. The numbers vary depending on the fortunes of world shipping trade, but at times there are over 20 vessels laid up in the river as this is one of the cheapest places in the world, with deep water, between 13m and 15m, and excellent shelter. Although somewhat incongruous these silent, waiting leviathans have a certain mournful fascination but do not be too distracted by them; the tide runs strongly, and the high shores and ships make the breeze very fluky and it is easy to get set on to the large moorings. In addition, midway along this reach is the King Harry Ferry, which runs on chains, providing a short cut for cars between Falmouth and St Mawes. Identical at both ends, ascertain which way it is running and always pass astern, and not too close.

Lamouth and **Cowlands Creeks** open to port, both attractive and wooded, and both drying almost as far as the moorings at their mouth, where an anchorage can be found, between them and the deep water. Small craft can explore further on the tide, as far as the peaceful hamlets of Coombe and Cowlands, renowned for their plum orchards, where bilge keelers can dry out on the foreshore. There is a public telephone at Cowlands. Roundwood Quay, where the creeks meet, is all National Trust property. A finely preserved granite quay once used for shipping local minerals, it has several pleasant walks leading away from it.

Opposite, the river turns sharply to the east off Tolverne Point with more large ships sometimes in midstream, and on the shore you will see the thatched Smugglers Cottage, a popular local spot, with a landing pontoon and a number of mooring buoys for customers. This corner has an interesting history as during the last war it was used as an assembly point for part of the American fleet of D-Day landing craft, and was even visited by General Eisenhower. For nearly a hundred years trip boats have run here from Falmouth for cream teas, and although not a pub it has a licensed restaurant where lunches, suppers and barbeques, weather permitting, are a regular feature. The fascinating collection of photos and mementoes in the cottage are well worth seeing, in particular the *Uganda* room, devoted to the famous cruise ship and Falklands veteran that was laid up in the Fal between 1985 and '86.

At the next junction, just above Tolverne, the rivers divide, the Fal fading away rather insignificantly into Ruan Creek which continues to the east, and the Truro River,

The upper Fal - peaceful anchorage off Ruan Creek

heading northwards. One of my favourite anchorages is within the entrance to Ruan Creek, just off the old ruined boathouse on the north side where there is about 2m at LW, but watch out for three enormous moorings blocks close to the shore which only begin to show after half tide. The south shore dries extensively; shoal draught boats can push further into the creek and dry out at LW, overlooked by dense woods along the edge of Lord Falmouth's estate. Beyond the first bend the creek dries completely but small craft and dinghies can explore it on the tide deep into the rural depths of Cornwall, as far as Ruan Lanihorne, 3½ miles away where there is an old quay for landing and the Kings Head pub within five minutes' walk.

Just upstream of Ruan Creek, there is a second Truro Harbour Authority visitors' pontoon, and it is also possible to anchor anywhere along this reach out of the main fairway, although it is busy with trip boats during the day and an anchor light is advisable at night. Beyond tiny Church Creek on the west bank where the ruined spire of Old Kea Church rises above the trees the Truro River becomes much shallower, and the Maggotty Bank extends across the river from the eastern shore, least depth 0.7m LAT, the channel very narrow between the west shore and the conical green buoy at the outer edge of the bank. Following the bend round to starboard, past Woodbury Point where there is an isolated white house, there is a deeper pool, with 6 Truro Harbour visitors' moorings, yellow, and marked with details of suitable size/weight. These too, are £5 a night.

In Malpas Reach, where there are many moorings, there is as little as 0.3m at LWS in places. The tide runs strongly, a good 2 knots at Springs, and anchoring is not easy. At Malpas, pronounced 'Mopus', the elevated houses overlook the river, and below them is the yard of Malpas Marine with a private landing pontoon. They can usually provide visitors' moorings at £5 a day; either phone ahead (Tel: 01872 71260) or to enquire berth temporarily on the end of their pontoon, but do not leave your boat unattended as this is also used by the ferries to Falmouth. Customers can leave their dinghies on the pontoon, and a shower/toilet, water and diesel in cans are also available. Close by there is a very small shop/Post Office, phone and the Heron Inn which does good pub food.

Truro

Tresillian River branches to starboard, and mostly dries. Truro River continues past Malpas, and is well worth exploring as far at Truro. Still used by coasters the channel is well-marked, but do not leave Malpas any earlier than three hours before HW Truro, and keep in midstream, leaving the conical green buoy and private landing pontoon off the private pontoon and housing development at Victoria Quay to starboard. A large bank extends west of the next long bend, so keep well towards the western shore before turning north past the first of two more conical green buoys which must be left to starboard. Beyond the second, the channel swings back to the eastern bank, high and wooded, past three port hand red cans, and then north-west again off Sunny Corner where there are always a number of boats laid-up on the beach. At the final red can, steer for the end of the Lighterage Quay, where there is a red beacon and the river narrows to about 100m wide. Coasters berth here, and at the northern end of the quay there is a flood prevention barrage with a flood gate that normally remains open except when higher than average tides are anticipated. The gate is 12m wide and poses no problems normally but make allowance for the fact that the tidal flow increases through this restriction. There is a waiting pontoon just downstream of the gate if it is closed or if you arrive too early on the tide to make Town Quay; beyond it, there is a launching slipway on the eastern, Boscawen Park, shore. Here, the river widens into a broad shallow reach, the Cathedral and houses of Truro now clearly in sight but the channel, marked by red posts with square red topmarks to port and green posts with triangular green topmarks to starboard, winds back to starboard past the playing fields, then close to the eastern bank before it swings back towards the long quay on the western bank. Following the line of this shore it narrows, with a large modern Tesco superstore to port, and the deepest water closest to the old warehouses to starboard. Around the bend at Town Quay the river divides into three cul-de-sacs – berth in the port hand one, adjacent to the Harbour Office.

There is water here for about 2 hours either side of HW and if you wish to remain longer you will dry out in soft mud, and it will cost £4.95 a night inc VAT for a 30 footer. There is a toilet and shower on the quay, fresh water, rubbish bins and a chemical toilet disposal point. It is perhaps not the most scenically memorable berth in the West Country but the attractive Cathedral City of Truro, within 5 minutes' walk, more than makes up for this. A major shopping centre which is able to provide all normal requirements. There is a chandler right on the quay, and another in New Bridge Street, many good pubs and restaurants and a main line railway station should you need to land or pick up crew.

PORT GUIDE: FALMOUTH
AREA TELEPHONE CODE: 01326

Harbour Masters: Falmouth: Captain David Banks, Harbour Office, 44 Arwenack Street, Falmouth TR11 3JQ (Tel: 312285 or 314379). Truro/Penryn: Captain Andy Brigden, Harbour

Office,Town Quay, Truro (Tel: 01872 72130 or 373352 for Penryn). St Mawes, Mr D Tatum, The Quay, St Mawes (Tel: 270553)

VHF: Falmouth, VHF channel 16, working 12. 0800-1700 daily, call sign *Falmouth Harbour Radio*. Truro, channel 16, working 12, call sign *Carrick One*

Mail drop: Harbour Office will hold mail, also RCYC, Falmouth Yacht Marina and Mylor Yacht Harbour. Truro Harbour Office

Emergency services: Lifeboat at Falmouth. Falmouth Coastguard

Anchorages: Off Custom House Quay, in harbour clear of moorings. Off St Mawes, St Just, off Restronguet, Loe Beach, and in upper reaches of river. Charges within Falmouth harbour

Moorings/berthing: FHC Visitors' Yacht Haven, North Quay, Falmouth 40 boats, max 12m LOA, up to 1.8m draught, pontoon island in harbour for up to 16 boats, max 12m LOA (both facilities summer only), and12 deep water visitors' moorings. RCYC 9 visitors' moorings off Club. Mylor Yacht Harbour Ltd, Mylor, Nr Falmouth (Tel: 372121). St Mawes SC (Tel: 270686). Malpas Marine, Malpas, Nr Truro (Tel: 01872 71260). Truro Harbour Authority, visitors pontoons above Turnaware and Ruan Creek, 6 visitors buoys at Malpas, drying berths alongside at Truro

Marina: Falmouth Yacht Marina, North Parade, Falmouth (Tel: 316620). 330 berths, up to 100 visitors. VHF channels 80 and 37, call sign *Falmouth Yacht Marina*

Charges: (For 30ft boat per night inc VAT) FHC Yacht Haven £8.40, pontoon island £7.68, moorings £6.92; at anchor £3.45. Special rates for over a week. RCYC moorings £7.00. Falmouth Marina £14.50. Mylor Yacht Harbour, alongside £12.92; moorings £7.63. Malpas Boats £5. Truro Harbour Authority pontoons/moorings £5.00. Truro Town Quay £4.95

Dinghy landings: Yacht Haven, North Quay. Fish Strand Steps, Prince of Wales Pier, RCYC. Mylor yacht harbour pontoon. St Mawes, slipway

Water taxi: None, but RCYC has boatman and launch,1000-1900 daily in season VHF channel 37 call sign *Club Launch*

Phones: Nearest to Yacht Haven in entrance to Chain Locker pub. RCYC. Falmouth Marina. By church at Mylor. Malpas

Doctor: (Tel: 317317)

Dentist: (Tel: 314702)

Hospital: Nearest casualty Treliske,Truro (Tel: 01872 74242)

Churches: All denominations

Local weather forecast: Visitors' Yacht Haven office. Falmouth Yacht Marina office

Fuel: Fuel berth on Visitors' Yacht Haven, summer only, petrol and diesel,0900-1800 (Tel: 01831 774118). *Falmouth Industry* fuel barge in harbour, diesel only Mon-Fri 0800-1630, Sat,0800-1200 Falmouth Marina, diesel only, 24 hours. Mylor Yacht Harbour diesel alongside, petrol in cans during season 0830-1730,1700 Sat

Paraffin: Cox and Co, The Moor, Falmouth

Gas: Calor/Gaz. The Bosun's Locker. Falmouth Marina. Mylor Chandlery, Mylor Yacht Harbour

Water: Yacht Haven, Falmouth Marina, Mylor Yacht Harbour, St Mawes, Malpas, Town Quay, Truro

Tourist information: The Moor, Falmouth (Tel: 312300). City Hall, Truro

Banks/cashpoints: All main banks in Falmouth and Truro, with cashpoints

Post Office: The Moor. Sub Post Office in newsagents close to Custom House Quay

Rubbish: Skip on North Quay. Bins at Falmouth Marina and Mylor Yacht Harbour, Turnaware visitors pontoon, Malpas, Town Quay, Truro

Showers/toilets: Yacht Haven Amenity building. RCYC. Falmouth Marina. Mylor Yacht Harbour. Pandora Inn. St Mawes SC. Malpas Boats. Town Quay, Truro. Chemical toilet disposal facility at Mylor, Truro Town Quay and Penryn Quay

Launderettes: Yacht Haven Amenity building. Falmouth Marina. Mylor Yacht Harbour. Pandora Inn

Provisions: All requirements in Falmouth, EC Weds but many shops open in season. Number of shops open on Sundays, inc Spar at Albany Road, ten minutes walk from harbour and North Parade Stores opposite Falmouth Marina entrance and Leos supermarket, 5 mins walk. Most provisions also at St Mawes, basics at Mylor Yacht harbour, seven days a week in season. All shops, Truro

Chandlers: Bosun's Locker, Upton Slip, Falmouth (Tel: 312414).The Boat Store, Falmouth Yacht Marina (Tel: 318314). Nautibits Used Boat Gear, Port Falmouth Boatyard. (Tel: 317474). The Boathouse, Commercial Road, Penryn (Tel: 374177). Monsons,(bonded stores) Commercial Road, Penryn (Tel: 373581) bonded stores. Mylor Chandlery, Mylor Yacht Harbour (Tel: 375482) seven days in season. Reg Langdon, New Bridge Street Truro (Tel: 01872 72668). Penrose Outdoors, Town Quay, Truro (Tel: 01872 70213) Millers Nautical Bookshop, Falmouth (Tel: 314542)

Admiralty Chart Agents/compass adjusters: Marine Instruments at Bosun's Locker, Falmouth (Tel: 312414)

Liferaft service/repair: Inflatable Boat Services, Penryn (Tel:377600)

Repairs/hauling: Falmouth Boat Construction Ltd, Little Falmouth Yacht Yard, Flushing, Nr Falmouth (Tel: 374309) slipways/hoist. Pendennis Shipyard, Falmouth Docks, 80 ton hoist (Tel: 211344). Port Falmouth Boat Yard, North Parade, Falmouth (Tel: 313248) slip/hoist (18 ton). Falmouth Marina Services, 30 ton hoist (Tel: 211691). William's Boatyard, Penryn (Tel: 373819). M&L Yachts Boatyard, Freemans Wharf, Penryn, 50 ton slip (Tel: 378253). Penryn Bridge Boatyard (Tel: 73322). Mylor Yacht Harbour (Tel: 372121) 25 ton hoist. Heard's Boatyard, Tregatreath, Nr Mylor, Falmouth (Tel: 374441). Freshwater Boatyard, St Mawes (Tel: 270443). Percuil Boatyard (Tel: 01872 58564). Pascoe's Boatyard, St Just in Roseland (Tel: 270269). Malpas Marine, Malpas (Tel: 01872 71260)

Marine engineers: Marine-Trak (Tel: 376588 or 314610). Falmouth Marina Services (Tel: 211691). Challenger Marine, Penryn (Tel: 376202) M&L Yachts, Penryn (Tel: 378253) Falmouth Boat Construction (Tel: 374309). Mylor Yacht Harbour (Tel: 372121). Robin Curnow, outboards/ Seagull agents, Penryn (Tel: 373438). S.Caddy, Penryn (Tel: 372682)

Electronic engineers: Mylor Marine Electronics, Mylor Yacht Harbour (Tel:374001). Sea-Com, Falmouth (Tel: 376565)

Sailmakers/repairs: Penrose, Upton Slip, Falmouth (Tel: 312705). Spargo, King & Bennett, The Boathouse, Commercial Road, Penryn (Tel: 372107). South-West Sails, Islington Wharf, Penryn (Tel: 375291)

Riggers: The Boathouse, Commercial Road, Penryn (Tel: 374177). Falmouth Boat Construction, Flushing (Tel: 374309). Mylor Yacht Harbour (Tel: 372121). Nautibits (Tel: 317474)

Transport: Branch line to main line connections to London and north at Truro. Daily coach connections with rest of country. Road connections to M5 via Plymouth. Newquay Airport, 45 mins

Car hire: Avis, at Marina (Tel: 211511)

Car parking: Several large car parks in Falmouth. Parking at Falmouth Yacht Marina and Mylor Yacht Harbour

Yacht Clubs: Royal Cornwall Yacht Club, Greenbank, Falmouth (Tel: 311105/312126). Flushing Sailing Club, New Quay, Flushing, Nr Falmouth (Tel: 374043). Mylor Yacht Club, Mylor Yacht Harbour, Nr Falmouth (Tel: 374391). Restronguet Sailing Club, Mylor Yacht Harbour, Nr Falmouth (Tel: 374536). St Mawes Sailing Club,1 The Quay, St Mawes (Tel: 270686)

Eating out: Many options, from Chinese to Indian, fish and chips to good à la carte. Many bistros

Things to do: Falmouth Maritime Museum, off Market Street, Falmouth. Pendennis and St Mawes Castles. Good safe beaches/walks. Ships & Castles leisure pool, Pendennis Head

Special events: Falmouth Regatta Week/ Falmouth Classics, second week in August. Life Boat Day/raft race mid-August. Gig and Working Boat racing throughout summer

HELFORD RIVER

Tides: HW Dover +0600. Range: MHWS 5.3m – MHWN 4.2m. MLWN 1.9m – MLWS 0.6m. Streams attain up to 2kts in river at Springs

Charts: BA: B147. Stanford: 13. Imray: Y57

Waypoint: Voose N Cardinal buoy 50°05′77N. 05°06′90W

Hazards: Gedges/August Rock, Car Croc Rock, Voose Rock (all unlit). Bar/shallows within river on north shore. Rough approach in strong east wind/ebb tide. Beware pot and net buoys

'W*hen the east wind blows up Helford river the shining waters become troubled and disturbed, and the little waves beat angrily on the sandy shores. The short seas break above the bar at ebbtide, and the waders fly inland to the mudflats, their wings skimming the surface, and calling to one another as they go. Only the gulls remain, wheeling and crying above the foam, diving in search of food, their grey feathers glistening with the salt spray.*

The long rollers of the Channel, travelling from beyond Lizard Point, follow hard upon the steep seas at the river mouth and mingling with surge and wash of deep sea water comes the brown tide, swollen with the last rains and brackish from the mud, bearing on its face dead twigs and straws, and strange forgotten things, leaves too early fallen, young birds and the buds of flowers.

The open roadstead is deserted, for an east wind makes an uneasy anchorage, and but for the few houses scattered here and there above Helford Passage, and the group of bungalows about Port Navas, the river would be the same as it was in a century now forgotten, in a time that has left few memories . . .'

So begins 'Frenchman's Creek' Daphne du Maurier's bestseller that made the name of the Helford familiar to all, the haunt of the heroine Dona, and hideaway for the Frenchman, and his ship *La Mouette*. Although the numbers of yachts have certainly increased in the roadstead, her words are very true, and large areas in this gem of a

river still remain untouched. Overshadowed by the busy harbour at Falmouth, commercially there was never any reason for the Helford to over-develop, it merely served the needs of the surrounding farms and communities, with a few small granite quays dotted along its banks. Apart from the scattered fishing hamlets, it remained '. . . *unvisited, the woods and hills untrodden, and all the drowsy beauty of midsummer that gives Helford river a strange enchantment, was never seen and never known.'*

The few people that did pass along this unpopulated, silent waterway into the depths of rural Cornwall, bounded by high shores, and deep, mysterious woods, were mostly heading for Gweek, right at the head of navigation. As the nearest access to the inland town of Helston this was the focal point of waterborne activity, with sailing coasters and barges slowly working their way up on the tide, bringing cargoes of coal, timber and lime, and taking away granite, tin and farm produce, until the advent of rail and motor transport, and gradual silting resulted in its decline at the turn of the century, and the river slipped back into obscurity.

The other reason for the remarkable natural preservation of the inner reaches is the Duchy of Cornwall oyster fisheries, reputedly dating back to Roman times, these extensive beds have always inhibited the spread of moorings and other commercial development. This and the very limited facilities mean that today the Helford River,

Approaching Helford from Falmouth the August Rock buoy and other boats indicate this well hidden entrance

even in season, can still have an almost deserted quality, a solitude that is increasingly rare, and, fortunately, cherished by those who come here to seek it.

Approaches

Just under four miles from Pendennis Head, the Helford is a regular jaunt across the bay for many Falmouth boats, perhaps pausing to anchor off a beach or for a quick run ashore to the pub, but as the afternoon draws in, most return home. Although the entrance is well hidden away to the south-west, the various other boats entering and leaving will give a good indication. As you close Rosemullion Head, 'August Rock', the green conical buoy to seaward of the Gedges rocks (drying 1.4m LAT) should be left on your

August Rock buoy and the Gedges

starboard hand before bearing round into the river mouth as it opens ahead, running due west. There are no further hazards except very close to the shore, and depths average between 3m and 4m in the entrance. Beating in, there can often be a noticeable funnelling effect within the mouth, but overall, the shelter inside is excellent in anything, except, as we have already learned, easterly winds, when this is a place to avoid. Not only does Falmouth Bay and the approaches kick up a very short steep sea, the swell within the moorings and anchorage, particularly with wind against tide makes for a lot of discomfort, and unless you can tuck yourself away in one of the creeks, you are much better off in the Fal.

On the southern shore in the approach to the Helford the hidden entrance to Gillan Creek is easily located by the distinctive hump of Dennis Head across its mouth. This can be a delightful spot in the right conditions, but is really only of interest to shallow draught boats, as it dries for the most part, and the only deep water within its mouth is almost totally taken up with moorings. Car Croc, a particularly nasty rock, drying 1m, sits almost in the middle of the entrance, marked by an east cardinal buoy, BYB, but be warned, it extends further to the south-east than might be imagined so give it a good berth, passing midway between it and the south shore when entering, but also beware of the rocks extending to seawards from Men-aver Point. Ideally, for a first visit, arrive just after LW when all the hazards are easy to see, and feel your way in on the tide, anchoring clear of the local moorings off the houses at

Flushing if you can, or go further into the creek to the picturesque hamlet and church at St Anthony, where you will dry out, well tucked away in this hidden corner. Here, on the shingly foreshore is the small yard of Sailaway St Anthony (Tel: 01326 231357) and it is possible that they might be able to provide a mooring. The densely wooded creek beyond the sandy spit is particularly attractive when the tide is in, perfect for a dinghy trip, or just a walk along the road that follows it inland.

Returning to the main river – in winds between north and west there are some good anchorages just within the entrance, tucked up in the bight along the northern side beyond Toll Point where there is an isolated pine tree. There is a shingly beach and small boathouse at Porth Saxon, providing a quiet anchorage with depths of 1.6m quite close to the shore, and a pleasant walk from the beach up over the headland to Mawnan Church, which is set among the trees on the cliff-top overlooking the entrance to the river. This is a particularly lovely spot, and it is not difficult to see why, among the gravestones, you can find those of two eminent yachtsmen, Claud Worth, the grandfather of modern cruising, and his son Tom who circumnavigated the world in 1953 aboard the Giles designed cutter, *Beyond*. His epitaph is particularly succinct – 'Tom Worth, Who Sailed Beyond'.

The anchorage off Porth Saxon lies west of Tolls Point

The cove overlooked by trees just east of the tiny hamlet of Durgan is another good anchorage, alternatively anchor off Durgan itself, clear of the moorings. This picturesque cluster of houses is partly owned by the National Trust, as is the valley running down to the village in which the 40 exotic acres of Glendurgan Gardens are situated, open to the public on Mondays, Wednesdays and Fridays. There are no facilities at Durgan except a phone.

The southern side of the entrance is less hospitable, fringed with rocks, and a couple of small coves. Locals claim that the large house here was the *Manderley* of Daphne du Maurier's 'Rebecca', not 'Menabilly' on Gribbin Head, and certainly, Ponsence Cove below it, with a tiny boathouse seems to fit the bill. One thing is certain, there is no doubt where the inspiration for 'Frenchman's Creek' came from – just over a mile upstream you can explore it for yourself.

Beyond it lurks the Voose, a drying rock that has snared

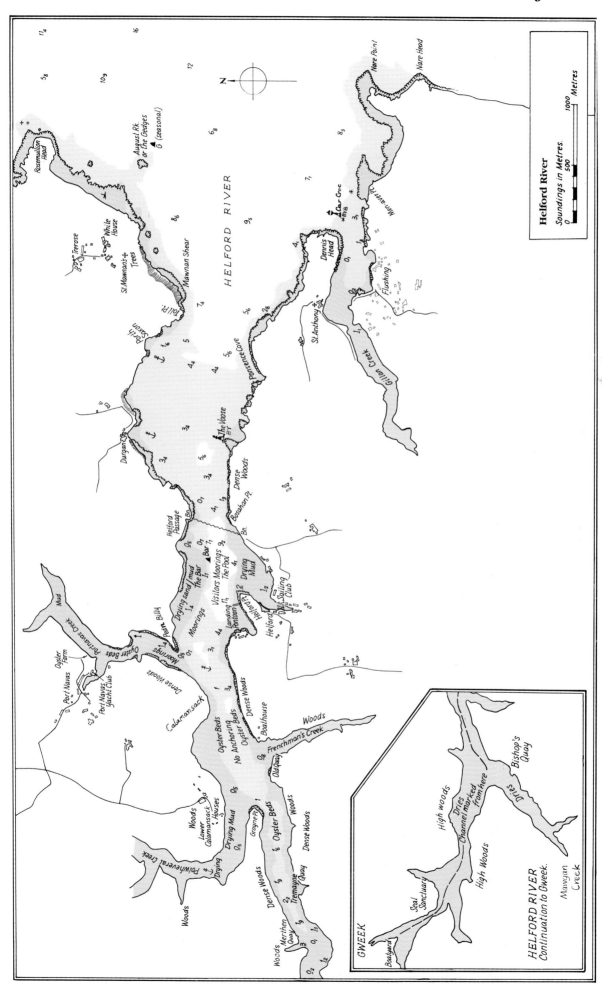

Helford River

Soundings in Metres.

0   500   1000   Metres

HELFORD RIVER
Continuation to Gweek.

GWEEK

103

a surprising number of people in spite of its north cardinal BY buoy. Approaching the narrows, if the tide is low, keep just over ½ a cable off the steep wooded shore leading up to Bosahan Point, and up towards the large concentration of moorings ahead, avoiding the northern, starboard, shore. Here, shallows extend right up to a cable along the shore with not much more than 0.5m LAT in places, right past the small boat moorings, beach, modern houses and pub at Helford Passage, as far as the green conical 'Bar' buoy. This is often not easy to spot amongst the surrounding boats, and inshore of it at LWS it dries extensively, a muddy bank which stretches as far as the entrance to Port Navas Creek, and is a popular local place for digging cockles. Being easy to reach by road from Falmouth, Helford Passage is the most commercialised part of the Helford River, centred around the Ferry Boat Inn which has a restaurant as well as bar food, and often live music. There is also a grocery/off licence, showers and launderette, telephone, and a passenger ferry across to Helford Point.

Moorings, anchorage, and landing

Containing the bulk of the Helford moorings, **The Pool** is nearly 15m deep in places, and averages about 6m, extending up the centre of the river. To port is the entrance to Helford Creek, but at LWS, this dries extensively, on a line from Bosahan Point to Helford Point, where there is a ferry slipway and landing pontoon.

In November 1884 the 'West Briton' newspaper revealed that *'. . . the beautiful Helford River has been visited this summer by an unusual number of yachts — as many as five having been at anchor there at any one time.'*

Plus ça change! Inevitably, today, nearly all the available space in the pool is taken up with moorings, and anchoring should not be attempted. However, there are a number of visitors' moorings available, green buoys or buoys with a green pick up buoy marked 'Visitor'. Either grab one, or contact the Moorings Officer, Jim Stephens, who might be listening on VHF channel 80 in his small office by the Helford landing pontoon, call sign *Moorings Officer* but, more often than not, will be out on the water in an 18 foot open white launch, ever ready to assist (Tel: 01326 221265 evening only). Remarkably there is still no charge for anchoring, and moorings are a very reasonable £4.50 a night.

You may drop people ashore at the ferry slipway, but

do not leave dinghies here. Instead, use the pontoon just upstream. This is private but the owner does not mind it being used as long as you leave a small donation towards its upkeep in the honesty box provided. The alternative, an hour or so after LW, is the landing pontoon off the Helford River Sailing Club, the impressive Scandinavian style wooden building among the trees on the eastern side of Helford Creek. The ferry will also run you ashore if you can hail it.

On the subject of voluntary donations, all the buoys in the Helford are privately maintained by the 'Helford River Navigational Aids Committee', a worthy cause run by Jim Stephens and local vet. John Head, and all yachtsmen who use the river are invited to contribute towards the upkeep of these vital aids. Should you wish to help, Jim will happily take your money!

The only feasible places to anchor are further upstream off Port Navas Creek, where depths reduce to about 2m LWS, clear of the moorings, or downstream of the Pool, towards Bosahan Point, but clear of the power cable crossing the river, marked by beacons on the shore. Here, an anchor light is advisable as local fishing boats come and go at night. In both places, you will be well away from the normal landing places and it is a good row if the tide is running hard, which it can do, up to 2 knots at springs on the ebb. Shoal draught boats can edge close inshore into Helford Creek, or off Helford Passage where they will dry out.

Facilities

The small village of Helford is not really evident at all from the river, just a few houses along the shore which disappears into the narrow creek, and normally it is best to pay the fee and land at the Helford Point pontoon to avoid the worry of the dinghy drying out. A short walk along the point brings you to the Shipwright's Arms, a classic thatched waterside pub, holding a strange attraction for thirsty crews, with a small restaurant, and a terrace outside. Winding on above the creekside quays and boathouses the narrow lane squeezes past thick walled stone cottages, whitewashed and covered in climbing roses, tiny gardens overflowing with flowers where you can sit and enjoy a cream tea, to the Post Office and general store, beyond which you will find a telephone.

At the head of the creek, a bridge and shallow ford lead to another row of equally picturesque cottages following the opposite side of the creek, where you will find the Riverside Restaurant, which, in such an out of the way place, survives on the sort of enviable reputation that enables it to lure its customers from far and wide. However, this is definitely one for a special occasion, prices are well out of the normal steak and chips range and reservations are a must.

That, essentially, is Helford village, and fortunately for its residents, cars are banished to a car park during the summer. You will soon reach it if you continue up the hill out of the village. A track leads down to the Helford River Sailing Club from the car park and the club is open daily except Mondays, with a good bar, food, showers and laundry available to visitors. When open, the club monitors VHF channel 80. (Tel: 01326 231460)

Helford Pool, looking west to Calamansack

Port Navas

Continuing upstream, just before you reach Pedn Billy point, and the entrance to Port Navas Creek, the large house with a small quay close to the water's edge is 'Bar', built by Claud Worth in 1928 for his retirement, and where he lived until his death in 1936.

Port Navas is another attractive wooded creek, although the eastern bank did not escape development and has a number of large houses overlooking it. However, just within the entrance, tucked away behind the point is the small, quiet pool at Abrahams Bosom, with just over 2m LWS, now mostly full of moorings, although shoal draught boats might find space to anchor close to the edge; if the wind turns to the east this is one of the few places where you will be completely sheltered. Most of the creek is shallow beyond the pool with extensive oyster beds, which are clearly marked by buoys, and a number of moorings, all of which dry at LWS. However, after half flood it is possible to get up to Port Navas Yacht Club which lies in a smaller creek on the port hand side. Keep in the centre and berth alongside the quay where there is a water tap, and the friendly Yacht Club welcomes visitors, with a bar and meals available during normal licensing hours. (Tel: 01326 40065). By arrangement, for a small charge, it is possible to dry out alongside for night.

Overhung with dense woods, Port Navas is a very peaceful little backwater, with just a few stone cottages, overlooking the narrow creek. The nearest provisions are at Constantine, a good half hour's walk away. It is, however, the home of the Duchy Oyster Farm, and following the road back towards the mouth of the creek from the Yacht Club you will find their buildings beside the main creek, where nearly 1½ million oysters are processed every year. Should a sudden extravagance overcome you it is possible to buy some, reflecting as they slither down as rapidly as your bank balance that these were once the staple diet of the poor. Mussels, too, are produced by the Farm, and *moules marinieres à la Port Navas* are probably a good bet for tonight. . .

The upper reaches

By far the most unspoiled area of the river lies beyond the great rounded woods at Calamansack, a clear stretch of water where there are no moorings because of the mussel buoys and oyster beds along both sides of the river. These are clearly marked by buoys and stakes, and anchoring is prohibited. The deeper water lies along the south shore, which is also heavily wooded with low, steep cliffs, but just before **Frenchman's Creek**, depths reduce considerably to little more than 1m at LWS, and half a mile further on, the river dries extensively. Yet again, this, and the other creeks are ideal for the dinghy, although on a reasonable flood, moderate draught boats can make it all the way to Gweek, which is accessible a couple of hours either side of HW.

It is impossible not be drawn into the romance of Frenchman's Creek, where '. . . *the trees still crowd thick and darkly to the water's edge, and the moss is succulent and green upon the little quay where Dona built her fire and looked across the flames at her lover. . . .*'. The reality is very much as it is described in the book, with glistening mud at low tide, where herons and oyster catchers roam. As the thin trickle of the flood creeps inland again, like the yachtsman in the book, *'the sound of the blades upon the water seeming overloud,'* you, too, can follow its winding course in your dinghy past blackened tree stumps emerging like creatures from the mud, to where the dense trees close in like a tunnel, brushing the incoming tide, and the silence becomes profound.

The atmosphere of Frenchman's Creek is undeniable; at Groyne Point, just a short distance upstream, **Polwheveral Creek** branches off to starboard, and although just as attractive, somehow has none of the mystery. If you are seeking some real solitude, round the bend beyond the moorings, shoal draught boats can find plenty of space to dry out.

Following the flood up to Gweek, the river passes between high wooded banks with mud fringed rocky shores most of the way and the deeper water lies in midstream. It is possible to land at Tremayne Quay, which is owned by the National Trust and there is a lovely woodland walk leading from it. Just downstream of Mawgan Creek the orange mooring buoy in the centre of the channel belongs to a local fishing boat, and beyond it the river dries completely at LW. Bishop's Quay on the south shore is private, and opposite the entrance to the creek a large bank fills the centre of the river, the channel swinging to port around it, but fortunately, from here on it is marked with posts; red with square topmarks to port, and green with triangular topmarks to starboard, which meander from one side of the river to the other until you reach the very narrow bottleneck with steep woods on either side just below Gweek. Once through the gap, the head of the river widens, the old coal quay lies to port, and more private quays to starboard, and, almost in the saltings, **Gweek Quay Boatyard** is straight ahead and it is possible to lie alongside and dry out overnight for a small charge. (Tel: 01326 221657)

This old, grass covered quay will always be a nostalgic place for me, for here, back in 1973, I found my own boat, *Temptress*, laid up and neglected, and spent many long happy weekends putting her to rights, and dreaming of where we would eventually sail. The atmosphere of Gweek has little changed, and it is still a relaxed and peaceful place, although the yard has grown considerably. Diesel and water are available, chandlery, a 30 ton crane and all normal yard facilities including engineer. In the small village close by there is a well-stocked grocery store, open on Sundays, Post Office, garage, the Black Swan pub, and a couple of cafes; however, the most unlikely thing that Gweek can provide, and one that children particularly adore, is the 'Seal Sanctuary', where injured and sick seals are looked after before being returned to the sea. It is on the north side of the creek, just downstream from the village, and open daily during the summer.

Passages
THE MANACLES
TO LAND'S END

**Passages – Manacles to Land's End
(and The Isles of Scilly)**
Charts for this section of coast are:
BA: 2565 Trevose Head to Dodman Point. 777 Land's End
to Falmouth. 154 Approaches to Falmouth 1148 Isles of
Scilly to Land's End. 2345 Plans in South-West Cornwall is
particularly useful as It has a large scale section covering
the Lizard
Imray: C6 Start Point to Lizard C7 Lizard to Scilly Isles
Stanford: 13 Start Point to Land's End and Padstow.
French: 2218 Du Cap Lizard à Trevose Head. 4812 Du
Cap Lizard à Start Point

Beacon	Range (M)	Frequency (kHz)	Call Sign
Marine beacons			
Lizard	70	284:5	LZ
Round Island	150	298.5	RR
Aero beacons			
Penzance heliport	15	333.0	PH
Coastal Radio			
Pendennis Radio, working channel 62.			
Land's End Radio, working channels 27, 88, 85, 64			

As the most southerly point of the British Isles, a major headland and tidal gate, the **Lizard** and the coast to the west of it into **Mount's Bay** and around **Land's End** has a justifiably notorious reputation and should always be treated with due respect and caution.

Composed for the greater part of precipitous granite cliffs – a firm favourite with the rock-climbing fraternity – this inhospitable and rugged coast is open to the prevailing winds; it is an area of strong tides, and invariably suffers from the long Atlantic ground swell, not to mention concentrated shipping and no absolute harbours of refuge.

But don't despair; in reasonably settled weather, with an experienced crew in a well-found boat it can be explored safely, but here, more so than ever, a wary eye should be kept on the weather, and options for shelter always close to mind.

Jack Pender, an old Mousehole fisherman, told me many years ago as I languished there in early September that *'west of the Lizard's no place for a small boat, come the end of August'*. I confess to a certain youthful panic and sailed that very night, scurrying back to Falmouth. But his words were born of a lifetime working these difficult waters, and myself a fisherman for several years, I often recalled them as we turned to run for home, the grey sea and sky piling to the west, white crests blowing before it, and the high, dim shore suddenly vanishing to leeward in the gathering gloom. Conditions can deteriorate in a matter of hours, and a contrary tide can produce large, often steep, and tumbling seas. It is not difficult to see how this particular coast has claimed so many ships and men over the years.

Bound west, take full advantage of the ebb tide out of **Falmouth Bay**, which starts to run to the south three hours before HW Dover (three hours after HW Falmouth) and once clear of the east cardinal BYB 'Manacle' buoy, a course of 220°T will pass all offshore dangers in the approaches to the Lizard, and should be held to a position three miles to the south of the headland to clear the race. Obviously, adequate allowance should also be made for the tidal stream setting to the west which begins about one hour before HW Dover. If bound round Land's End, because of the division of the streams into the Irish Sea and English Channel, a favourable tide can now be carried for nearly eight hours. Conversely, the passage east is not so obliging; a vessel carrying a fair tide down the North Cornish coast and round Land's End will invariably run into a foul tide off the Lizard, as from the turn of the tide to the south at the Longships, five hours before HW Dover, only three hours of favourable tide can be carried across Mount's Bay; once round the Lizard the ebb starting from Falmouth Bay will be gathering in strength against you. Bound east from Mount's Bay to Falmouth it is, therefore, advisable to arrive at the Lizard at slack water just over three and a half hours after HW Dover.

Hopefully, you will be sailing this spectacular section of coast in daylight and good visibility when it can be appreciated to the full. In favourable weather and settled

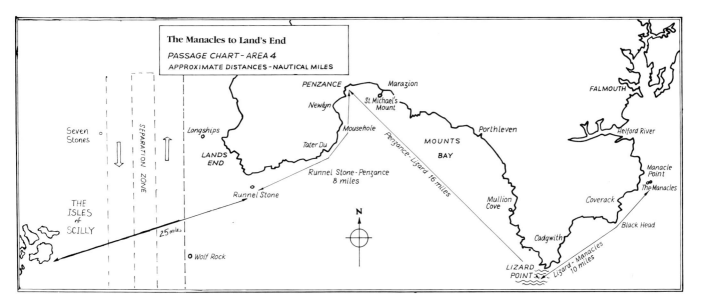

The Manacles to Land's End
PASSAGE CHART - AREA 4
APPROXIMATE DISTANCES - NAUTICAL MILES

conditions the coastline can for the most part be followed much closer inshore.

From Lowland Point, the land rises to a distinctively flat topped profile, steep-to and rocky with cliffs between 30 and 50 metres in height, and few off-lying dangers. It is interspersed with many attractive coves and bays, so if time permits, or if waiting for the tide, some of the smaller havens of the Lizard peninsular can provide good temporary anchorages and an interesting diversion.

Coverack, the first of these, is easily located by the conspicuous large hotel just south of the clustered houses of the village. **Black Head,** a mile to the south, is the next prominent feature and can easily be mistaken as Bass Point on the Lizard when approaching from the east in poor visibility. The tidal streams can run up to 3 knots at Springs, and in strong southerly winds considerable steep breaking seas will be encountered with wind against tide in the vicinity. If running for Falmouth in such conditions, lay a course several miles to seaward of both Black Head and the Manacle buoy, where similar poor conditions can also be encountered.

Beyond Black Head the coast falls back to form a wide bay, well sheltered from the west and a one time favourite haunt of the Falmouth Pilot vessels and quay punts waiting for business, and also of a Falmouth tailor's cutter, waiting to put a man aboard homeward bound ships so that the crews could walk ashore in a brand new suit. In its centre **Cadgwith** nestles in a rocky cove, the houses wedged tightly at the mouth of a narrow valley; **Parn Voose Cove** and **Church Cove**, with a small landing slip lie about a mile to the south, and the present Lizard/Cadgwith lifeboat station is spectacularly sited close by in a narrow crevice in the high cliffs at **Kilcobben Cove**. However, special care is needed sailing inshore along this section of the coast and the large scale Admiralty Chart No 2345 is a must. The Craggan Rocks, with less than 2m over them, lie just over half a mile SSE of Cadgwith, and the Voge Rock, covered by only 1.6m, lies two cables east of Church Cove. The yellow buoy further offshore marks the end of a sewer outfall. Bass Point is steep and topped by the old coastguard station with the distinctive white building of the former Lloyds signal station close by.

The Spernan Shoals and notorious Vrogue Rock, covered by 1.8m, lies four cables ESE of Bass Point, its position indicated by transit beacons ashore, and the passage between the rock and the shore should only be used in settled conditions. Normally, approaching from Black Head without any diversions inshore, your course should be laid well to seaward of Bass Point. With rocks extending over half a mile to the south of the Lizard, and overfalls, severe with wind against tide, it is advisable to give the whole area a good berth in anything but settled conditions, standing off between two and three miles. In rough weather five miles is not unrealistic, for the seas inshore can be very confused.

Tidal streams are strong, the west-going ebb can run at over 3 knots at Springs, the flood slightly less, and it goes without saying that a fair tide is essential for a sailing vessel with limited power. In westerly winds its full force will not be felt until clear of the Lizard, and after spells of weather from that quarter, a considerable ground swell will be encountered.

It is a headland to approach with extreme caution in poor weather or bad visibility; the vast list of vessels lost in the vicinity over the years are an adequate testimony to its natural dangers. Care should also be taken to note shipping movements as they concentrate towards Land's End, the activities of fishing boats, and, yet again, beware of poorly marked pot and net buoys, often without flags, and half-submerged in the tide. One more unusual 'hazard' that certainly added a few grey hairs on one occasion, was the sudden appearance of a totally unmarked rock nearly three miles south of the point, black, and awash with breaking waves, right ahead. The chart confirmed absolutely no existence of such a horror; several minutes of dry-mouthed panic elapsed before it dawned on me that it was an enormous basking shark, a plankton eating summer visitor to Cornish waters and, otherwise, totally harmless.

The **Lizard lighthouse** is a prominent and distinctive long white building with two octagonal towers, the easternmost topped by the light, its five million candlepower

The Manacles to Land's End

The Lizard lighthouse

giving a 25 mile range (Fl 3s). The first light on the headland, and the first in Cornwall, it was established in 1619 by Sir John Killigrew, amidst much local protest at the adverse effects on the profits from the wrecking, and Killigrew himself eventually abandoned his light in favour of the more lucrative spoils from the sea. Several other attempts to provide a lighthouse on the headland during the 1700s ensued, notably Fonnereau's twin towers with a fire in each and a cottage between in which an overlooker lay on a couch watching for any relaxation of the firemen's efforts, a blast on a cowhorn *'awakening them, and recalling them to their duty!'* The onset of oil lighting in 1813 put an end to such navigational uncertainty and by the end of the century both an electric light and a foghorn had been established, with a correspondingly dramatic decline in the loss of shipping.

The Boa, a rocky shoal 2½ miles west of Lizard point is the last offshore hazard in the vicinity and although well covered with over 20m it creates a lot of overfalls, and even breaking seas in south-westerly gales. If the Lizard has been given a berth of 3 miles as recommended, heading into Mount's Bay you should pass clear to the south of the Boa. Its location is usually easy to spot from the concentration of pot buoys.

The western side of the Lizard, exposed to the full force of the Atlantic gales is high, rugged and spectacular for the next five miles as it bears north-west into **Mount's Bay**. The tall jagged pyramid of Gull Rock and Asparagus Island enclosing the beauty spot of **Kynance Cove** are a distinctive feature, and both Rill Point and Predannack Head should be given a reasonable berth. **Mullion Island** will begin to open as Predannack Head is passed, and in favourable easterly conditions, an anchorage can be found off the small harbour of **Porth Mellin**. Although a passage exists between the island and the mainland, it is not recommended, and the approach is best made to the north of the island.

The character of the coast begins to change considerably as Mount's Bay is entered further. Beyond Pedngwinian Point, the high cliffs recede, and the long sand and shingle beach of Gunwalloe and Loe Bar stretches away northwards. Particularly vulnerable to southerly gales, many vessels struggling to escape round the Lizard from Mount's Bay have come to grief on **Loe Bar**, including the frigate

HMS Anson in 1807 when 100 men were lost in the surf trying to reach the shore. Witnessing the catastrophe, Henry Trengrouse was inspired to invent the rocket line throwing apparatus which is still used by the Coastguards today.

Porthleven, lies at the northern end of Loe Bar, with a conspicuous clock tower by the harbour mouth and should only be approached in offshore winds and settled conditions, and from here the northern shore of Mount's Bay begins to trend more to the west. Welloe Rock, drying 0.8m, lies three miles due west of Porthleven, and the Mountamopus shoal, 1.8m LAT is a mile south-west of Cudden Point. Passing to the south of the south cardinal buoy marking this hazard, the distinctive pyramid of **St Michael's Mount** topped by a spire and turrets is unmistakeable, and keeping it on the starboard bow, **Penzance**, 2½ miles to the west is easy to locate, with no further hazards except the 'Gear', a drying rock marked by a black and red beacon, just under half a mile due south of the harbour entrance. Pass to seaward of the Gear if bound across the bay to **Newlyn**.

Continuing west from Mount's Bay, if leaving from Newlyn, a course can be laid just over a cable from the shore inside Carn Base and Low Lee shoals, the latter marked by an east cardinal buoy, and past Penlee Point with its old lifeboat house and slipway. This, and the memorial garden beside it, remain as a sad reminder of the tragic loss of the lifeboat *Solomon Browne* with its entire crew of eight on the 19 December 1981, while attending the wreck of the coaster *Union Star* near Lamorna Cove. It was the last launch from the slipway, and the replacement Penlee lifeboat is now kept afloat in Newlyn harbour.

Once past **St Clement's Isle** and **Mousehole** the impressive grass topped granite cliffs form a continuous line, broken only by a few tiny coves and sandy beaches such as Penberth and Porthcurno, and finally the section between **Gwennap Head** and **Land's End** is particularly precipitous. Stay half a mile offshore until abeam of **Tater Du** light, when a course is best laid directly to pass just south of the large 'Runnel Stone' YB south cardinal buoy, which marks the outer end of a rocky ledge extending nearly a mile southwards from Gwennap Head, with the drying Runnel Stone at its extremity, a considerable hazard to shipping, and the scene of many wrecks. It is not unknown for the Runnel Stone buoy to break adrift in the

Tater Du lighthouse, Gwennap Head beyond

heavy seas that often run along this most exposed corner. There are, fortunately, two conical beacons inshore on Hella Point, the outer red, the inner black with a lower white band, providing a transit of 352°T over the position of the rocks. When the white base of the inner beacon is also visible above the cliff top you will pass safely to the south of the hazard.

This is, in any circumstances, an area to be navigated with extreme caution. Although the tides in Mount's Bay are weak they rapidly gather in strength towards the Runnel Stone, probably attaining 5 knots at Springs, becoming increasingly unpredictable as the main tidal streams divide into the Irish Sea and English Channel. As mentioned earlier, this is a very unfairly balanced tidal gate; the north-west stream begins three hours before HW Dover and runs for nearly 9½ hours, the east-going stream six hours before HW Dover lasting for a mere three. The Atlantic ground swell is rarely non-existent for the passage 'around the Land' to the north Cornish coast and this, and the coast beyond, is an area to take very seriously.

A favourable forecast is essential to attempt Land's End, or the passage to the Isles of Scilly which is covered in detail elsewhere in this book. Bound round Land's End from Mount's Bay, if you leave about one hour before HW Dover you will have a favourable tide for the next seven hours and, hopefully, soon be well on your way.

At night, this passage area is well lit, although once south of the Manacle buoy (Q (3) 10s) steering 220°T the Lizard light (Fl 3s 25M) will be obscured for the next five miles until you are a couple of miles south-east of Black Head. Inshore, Coverack and Cadgwith show only small clusters of lights. Once clear to the south of the Lizard, Tater Du lighthouse (Fl (3) 15s 23M) will be seen away to the north-west on the far side of Mount's Bay, and in the far distance to the west, Wolf Rock lighthouse (Fl 15s 23M).

Shipping bound to and from Land's End converges on the Lizard and this can be a busy stretch of water; many fishing boats are likely to be encountered, particularly trawling at night and their movements should be carefully observed and avoided. There are also regular Naval exercises in the area.

Entering Mounts Bay, the whole of the northern shore appears as a continuous mass of lights from Marazion to Newlyn, and it is worth noting that both the Lizard and Tater Du lights become obscured in the approach to Penzance, although by then the lighthouse on the south pier (Fl WR 5s 21/17M) should easily be visible, along with the harbour light at Newlyn (Fl 5s 9M). 'Low Lee' (Q (3) 10s) is the only lighted buoy in Mount's Bay, one mile south-east of Newlyn harbour entrance.

Heading west to Land's End beyond Tater Du, the Runnel Stone buoy (Q (6) + L Fl 15s) lies within the red sector of the Longships lighthouse (Iso WR 10s 19/18M) and is also covered by a (FR) light showing 060°T to 074°T from Tater Du, and it is wisest to stand on past the buoy to the westwards until the white sector of the Longships opens before altering to the north-west. Pass well to seawards of this light, the tide in its vicinity runs hard and unpredictably. You are now in the Land's End inshore traffic zone and a lot more shipping is likely to be encountered.

In poor visibility additional aids to navigation are the fog signals at the Lizard (Siren Mo (N) 60s), Wolf Rock (Horn 30s), and Longships (Horn 10s); the 'Manacle' buoy has a bell, the 'Runnel Stone' buoy a whistle and bell, and Newlyn harbour entrance (Siren 60s).

Land's End

COVERACK AND CADGWITH

Tides: HW Dover +0600
Charts: BA: *777*. Stanford: 13. Imray: C6, C7
Hazards: Coverack: Small boat harbour dries completely. Manacles. Isolated rocks off Dolor Point (unlit) and to NW of harbour.
                      Fine weather anchorage only and dangerous in onshore wind
                      Cadgwith: Bow rock (unlit) in entrance. Fine weather anchorage only and dangerous in onshore wind

Between The Manacles and Black Head, **Coverack** is a small, picturesque drying harbour popular with tourists, and still the base for a small fleet of fishing boats, open craft used for potting and handlining and known locally as 'cove boats' or 'toshers'.

In settled weather with offshore wind and an absence of ground swell, the bay makes a tenable anchorage although with any indication of a shift of wind to the south or east it is no place to linger.

Approaches

Approaching, the conspicuous large hotel to the south of the village is the easiest landmark to spot, and with the Guthen Rocks off Chynalls Point to the south, and the Dava Rocks extending nearly half a mile from Lowland point to the north end of the bay do not attempt to cut corners. Enter on a westerly course, sounding your way in to the best anchorage in about 3m, a cable or so NNE of the pierhead. The harbour, enclosed by a small, sturdy granite wall, dries to beyond the entrance with a clean, sandy bottom but the shore immediately to the west is rocky and very foul. Although there is a minimum depth of 2.4m alongside the pier end at MHWN, due to the crowding of the local boats which lie on heavy rope fore and aft moorings drying out is not really a viable proposition. Far better to anchor off and dinghy in, landing at the ladders on the pier or the large slipway across the head of the harbour.

Coverack has a small harbour which dries

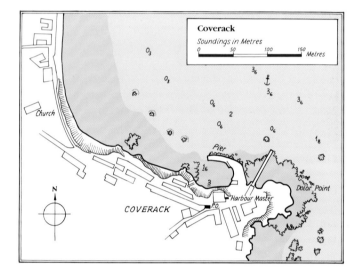

Coverack, like so many other Cornish harbours, grew with the extensive pilchard fishery, and the old salt store overlooking the harbour is now, inevitably, a gift shop. All the basic facilities are available; provisions, Post Office, several cafes, pub/hotels, including the curiously named Paris Inn, nothing to do with the EC or 'entente cordiale', but named after the *City of Paris*, an American liner stranded nearby in 1899 when her 700 passengers were brought safely to shore by local boats, before she was eventually refloated. Early closing Tuesday. Water is available from the Harbour Master but the nearest fuel is at a garage nearly two miles away. The Old Lifeboat house is also a gift shop but an inshore lifeboat is still stationed here.

CADGWITH

'There is a bench from which the whole of the bay can be seen where the fishermen sit in patience, and scarcely turn their eyes from the sea. It really is very exciting to hear the pilchard cry for the first time; visitors rise up and leave their dinner and amusements, and every man who dwells in the village, whatever he may be doing, is called by a strange and terrible cry to come and help in the 'take'. The boats are always ready in the bay, but the real time to see the pilchard take is by moonlight, when the fishes look like living silver . . .'

Cadgwith, it would seem, was popular even as far back as 1885 when this was written and although the huge shoals of pilchards vanished in the 1930s, the holidaymak-

Cadgwith can be an interesting stop in settled offshore weather

ers still descend in their droves. Nevertheless, it is still very much a working fishing community, and the traditional atmosphere of the village lingers in the tight cluster of cottages nestling snugly in a steep green valley at the head of a tiny cove. Almost too photogenic, white painted walls and heavy encrusted thatches contrast sharply with the more familiar granite cottages, and the curious dark green serpentine rock that is local to the Lizard. There is no harbour, and the sizeable fleet of small boats is hauled up the shingle on wooden rollers, a daily spectacle much enjoyed by the visitors. Thumping softly, an ancient and magnificent single cylinder Hornsby donkey engine powers the winch; in the cool, dark cellar where it lurks, the walls are covered with faded paintings, curling sepia photographs, and the sweet aroma of warm oil pervades.

Approaches
Well sheltered from the west, the cove should only be approached in settled weather and offshore winds, and is not recommended for an overnight stop. The houses will be spotted from offshore, forming a break in the flat line of the cliffs, and closing them you will find what is effectively two coves, split in the middle by a low rocky outcrop called the Todden. The extension of this, a group of rocks – the Mare – extends to seaward, and it is advisable to sound in and anchor off them in about 2-3m. Do not proceed any further into the cove as the Bow is a dangerous rock right in the centre which covers at quarter tide, and the entrance is also in frequent use by the local boats. Landing is easy anywhere on the shingle beach, but make sure you take your dinghy well clear, to avoid obstructing the winching operations.

There are limited provisions, a Post Office and pub/hotel/cafe, and a pleasant walk up on to the cliffs overlooking the cove, past some lovely cottages to a small black hut, high on the headland. It was here that those fishermen used to sit waiting and watching, and once the cry of 'Hevva, Hevva!' sent them to sea, from this high vantage point the 'huers' would direct the boats towards the shoals by a series of special hand signals and wild shouts through a large tin speaking trumpet.

Huers' huts can still be seen in many places along the Cornish coast, abandoned after the strange demise of the pilchard and the herring. The huge shoals that were once the livelihood of so many small villages like Cadgwith began to dwindle mysteriously in the mid 1930s, and after the war they were never found again, a phenomenon that has never really been explained. Today, the fleet of cove boats works nets, pots and handlines, rock hopping along this beautiful but at times very wild stretch of coast.

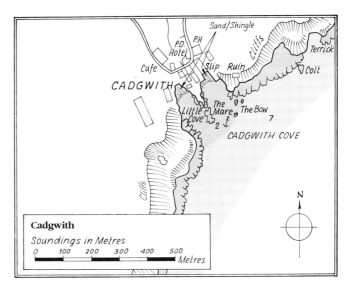

111

MULLION AND PORTHLEVEN

Tides: HW Dover +0555
Charts: BA: 2345, 777. Stanford: 13. Imray: C7
Hazards: Mullion: Mullion Island, many rocks close inshore. Fine weather anchorage only and dangerous in onshore wind
Porthleven: Harbour dries. Deazle rocks, Little and Great Trigg rocks (all unlit) to west and east of entrance
Dangerous to approach in onshore winds

This small drying harbour three miles north-west of the Lizard, spectacularly situated in a magnificent stretch of coast, provides another interesting stop in suitable offshore winds and settled weather, and is also convenient if waiting for a fair tide eastwards round the Lizard.

Approaches

The approach is straightforward: both Mullion Island and the conspicuous Mullion Cove Hotel high on the cliffs above are easy to see. Although there is a narrow channel between the island and the mainland used by local boats it is not recommended and the northern end of the island should be given a good berth, entering the anchorage midway between it and the small Porth Mellin harbour wall. This was built in 1895, a somewhat lethargic response to the disastrous loss of the cove's entire fishing fleet fifty years earlier during a freak storm. Today there are only a few local small boats based here, for the years have done nothing to change its exposure to the prevailing wind and sea.

The harbour is now owned by the National Trust and from May to September there is a resident Harbour Master. Although there is a berth alongside the western wall (as long as it is not being used by the fishing boats) it is not recommended as there is frequently much surge in the harbour which is, incidentally, closed to fin keel boats, and no overnight stays are permitted in any craft.

The anchorage between the harbour and Mullion Island is, therefore, the only option, well sheltered to the north and east with good holding on a sandy bottom in 7m

Porth Mellin harbour. The best anchorage lies between the outer wall and Mullion Island

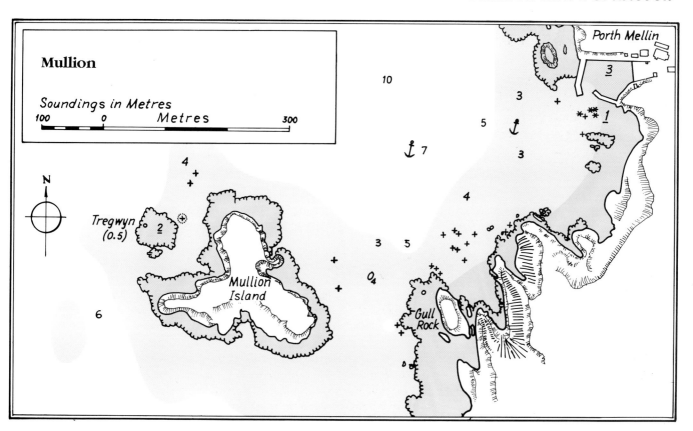

midway between the harbour entrance and the northern end of the island. This is a nature reserve and landing is not allowed.

An unusually remote anchorage, well off the normal track, it has not always been so . . . Writing in September 1868, R T McMullen, anchored here in a north-easterly gale aboard his *Orion*, counted sixty-four vessels sheltering, and commented, '. . . *I was surprised to see how regularly they were arranged according to their ability to work offshore if the wind were to fly in. The* Orion *was first in line with three pilot cutters, then came the sloops and yawls, and a brig-rigged steamship. Next schooners and ketches, then brigs and barks; those in the first division were almost still on the water, the second were rolling perceptibly, the third decidedly uneasy, and the last, having no protection at all from Mullion Island, were rolling miserably .'*

Facilities
Dinghy ashore when the tide allows and land on the slipway at the head of the harbour, where there are just a few houses, a cafe, gift shop, telephone and public WC. Off the main tourist track it is a particularly unspoiled little corner with some magnificent views and excellent walks along the cliffs in both directions. Most normal provisions are available at Mullion village, an uphill walk of just over a mile, and the Mullion Cove Hotel is open to non-residents with meals available in the bar. Owing to the potentially exposed nature of the anchorage I would not, however, recommend leaving your boat unattended for any length of time, and to leave with any hint of a wind shift to the south.

PORTHLEVEN
I always feel rather sorry for Porthleven, for it is certainly a magnificent example of the harbour builder's art, a long protective entrance leading to a fine basin enclosed by massive granite walls. However, its history reveals little commercial success, a chapter of disasters, starting right from the beginning when a group of speculators obtained an Act of Parliament ostensibly to build a harbour of refuge in 1811. Disagreement and squabbling dogged the venture and the harbour was not completed until six years later, surviving for another six before it was devastated by a storm, a fact that did little to encourage shipowners to use it. Although it was rebuilt and greatly improved in the mid-1800s, its fundamental failing was the fact that it dries completely and the narrow entrance faces right into the prevailing south-westerlies, rendering it frequently unapproachable, and forcing its closure with large baulks of timber in heavy weather.

Today, Porthleven is mostly used by pleasure craft, and a small but diminishing fleet of fishing boats. However, as with all the other small harbours of the Lizard, with offshore winds and settled conditions it makes another interesting place to visit off the regular cruising track, and a feasible overnight stop providing you don't mind drying out.

Approaches
The harbour mouth lies at the northern end of Loe Bar, a long shingle beach that begins just over two miles north of Mullion Island, the houses on the hillside easily visible on the cliffs, and the south pier, with a prominent clock tower at the landward end should be closed on a bearing of 045°T. The harbour dries right to the entrance at MLWS, but is accessible sensibly from half-tide onwards, and if waiting for water, anchor in 5m, 200m due south of the pier. Care must be taken to avoid the submerged Little Trigg rocks off the pier end though there are no further

Porthleven dries completely at LW but is well protected once inside

hazards once past the old lifeboat house on the north shore. Pass through the outer entrance and into the inner basin where the local boats mostly lie on fore and aft moorings with heavy ground chains running north-south up the harbour. Visitors should berth inside the entrance on the East Quay and seek out the Harbour Master Dennis Swire if he has not already spotted you. His office is well marked on the East Quay and he will advise on the most suitable berth. Overnight charge for a 30 footer is £4.70.

There is no VHF watch, and vessels over 40 feet should make prior arrangements by phone. (Tel: 01326 561141) The south quay with a crane at its end for lowering the timber baulks is used by the local fishing boats and should be avoided.

Facilities

A popular place with holidaymakers, Porthleven provides a curious contrast of old fishermen's cottages and converted net stores, and a long row of typically bold Victorian semi-detached houses that completely dominate the entrance. All normal facilities are available: Post Office, banks, launderette, and water is available in cans from the Harbour Master or 'Cowls' garage at the north-west end of the harbour. As well as fuel in cans this also provides a rare treat for lovers of old marine engines, like Kelvins and Brits, as the garage is a virtual shrine to them. There are cafes, pubs and restaurants, including the inevitable 'Ship Inn' overlooking the harbour entrance, and an interesting walk can be had along the great shingle sweep of Loe Bar as far as Loe Pool, a large freshwater lake, formed when the bar sealed off the estuary of the River Cober that was once navigable as far inland as Helston. It is part of the large Penrose Estate, and now owned by the National Trust. There are some delightful walks through the woods along its banks.

Porthleven

Soundings in Metres.

100 0 Metres 200

Garage (Fuel) (Water)

Inner Harbour 3 2

Harbour Master

Ship Inn 2

F.G. (occas)

Great Trigg Rocks

F.G.10m 4M
4 (occas)

Little Trigg Rocks

N

PENZANCE

Tides: HW Dover +0550. Range: MHWS 5.6m — MHWN 4.4m MLWN 2.0m — MLWS 0.7m. Tidal dock gate manned every tide from 2 hours before to 1 hour after HW
Charts: BA: 2345. Stanford: 13. Imray: C7
Waypoint: South Pier Head 50°07'02N. 05°31'62W
Hazards: Gear Rock (unlit). Outer harbour dries, tidal wet dock. Harbour approach very dangerous in strong southerly weather. Pot and net buoys in Mount's Bay. Keep clear if *RMS Scillonian* or other commercial shipping entering or leaving

W hen McMullen put into Penzance in 1868 it was still in its busy commercial heyday and he moaned bitterly about the state of the quays, '. . .*which are allowed by the Corporation to be in so offensive a state, encumbered with coal dust that nothing short of real distress will drive me into the nasty harbour again.*'

The coal has long vanished from the quays of Penzance but so, too, has most of the waterborne trade. Now a busy centre for tourism, the town grew around the export of tin, reaching its peak in the mid 1800s when nearly half of the minerals mined in Cornwall passed through the port, stacked in 300lb ingots on the quayside for shipment to places as far afield as Russia and Italy. A major centre for the export of salt herring and mackerel, there are records

of a thriving trading and fishing village here as early as 1300, but the major extensions to form the present day harbour were made in 1745-72 when the Albert Pier was built, with further improvements during the nineteenth century. The dry dock was built in 1814, and is still run by Holman's. Penzance was also home to the first lifeboat in Cornwall in 1803, but this was discontinued in 1917, and the lifeboat is now based in Newlyn. Until the late 1980s this was the westernmost Trinity House buoy depot and today it is a museum, 'The National Lighthouse Centre' and open daily during the summer.

The outer harbour dries almost completely at LAT and is given over to a large number of local moorings on fore and aft trots. Visitors normally use the tidal wet dock, entered

Yachts must raft in Penzance inner harbour where they will lie afloat and secure in all weather

115

through a hydraulic gate, and only accessible for **two hours before HW until one hour after HW.** Here there is ample berthing for up to fifty yachts, afloat at all times in total protection. A small amount of commercial traffic still uses the basin, and it has become popular for refits and repair work. It is, perhaps, not the most attractive setting, but the warm welcome extended by the dock staff and Harbour Master, Captain Martin Tregoning, more than makes up for the surroundings. A keen yachtsman himself, who sails a 28 foot gaff cutter, he has done much to encourage visitors and this is certainly reflected by their increasing numbers. By arrangement, boats can be left unattended without worry for a few days or longer, making it an ideal base to explore west Cornwall, and the rugged Land's End peninsula with good bus services and cars for hire.

Approaches

Approaching from the east, pass to the south of the south cardinal 'Mountamopus' buoy marking a 1.8m shoal, holding the unmistakeable bulk of **St Michael's Mount** on your starboard bow. The conspicuous tower of St Mary's Church provides a good landmark to locate Penzance. Gear Rock, which dries 1.8m, lies just over 800m almost due south of the harbour entrance and is marked by an unlit isolated danger beacon, not particularly hazardous when arriving from the east but right on the approach from the southwest. Penzance Harbour lighthouse, a white tower, displays a red and white sectored light (Fl WR 5s, red sector through 159°-268°T 17M, white sector through 268°-345°T 21M). The white sector safely clears Gear Rock, and if approaching from the 'Low Lee' buoy steer a north-easterly course until the white sector opens. There are lights displayed vertically from the Harbour Master's Office by the north side of the dock gate indicating whether it is open or closed: red over green = closed; red over red = open.

Accessible with safety in winds from south-west to north, *Penzance should be not considered as a harbour of refuge in bad weather,* particularly with winds from the east through to south, when heavy breaking seas can build up in the shallowing approaches, often breaking over the south pier, and making the whole entrance highly dangerous. In these conditions Newlyn is the only place to consider, approaching with extreme caution and as near to HW as possible. However, in reasonable weather, and offshore winds a temporary small craft anchorage can be found just over 350m NNE of the end of the Albert Pier, in about 1.2m LAT, or further offshore depending on your draught, ensuring that you are clear of the fairway into the harbour entrance, or alternatively anchor 350m due south of the south pier head.

The Harbour Office monitors VHF Channel 16 during normal working hours, and works on 12, call sign *Penzance Harbour Radio.* It is also manned HW-2 to HW+1 when the tidal dock is open. Although the outer harbour is full of moorings, it is possible to dry out alongside the Albert Pier with the Harbour Master's permission.

Waiting for the tide, the alternative to anchoring off is lying on the inside of the South Pier in the berth used by the Isles of Scilly ferry *RMS Scillonian III* which is

One can lie on the pier if waiting to lock into Penzance

normally empty Monday to Friday from 0930 to 1830, and summer Saturdays from 1345 to 1830. At low water keep close to the south pier as the water shoals rapidly on the northern side of the entrance, but between the convenient ladder half way along the wall and the lighthouse a depth of 1.8m will be found even at LAT. You can lie alongside the large floating fenders – be warned, though, if you clamber onto them they roll instantly, and you'll end up swimming. Check in with the Harbour Master once berthed and do not leave the boat totally unattended or obstruct the stone steps by the entrance to the wet basin as this is in constant use by local fishing trip boats.

The dock gate is manned two hours before HW every tide, day and night, and opens soon afterwards. Once in the basin, the Berthing Master will allocate a berth alongside, overnight charge for a 30 footer £8.10 inc VAT, with every 3rd night free. You will be given the code to the locked toilets and showers situated beneath the Harbour office in the end of the large white building nearest to the lock gate. The public toilets on the south pier should only be considered in the direst of emergencies! The Berthing Master can also arrange for water, small quantities of diesel and Calor/Gaz refills; petrol is available in cans from the garage near the railway station.

Facilities

Penzance is a popular tourist spot, a busy and interesting old market town with many fine buildings. This is the

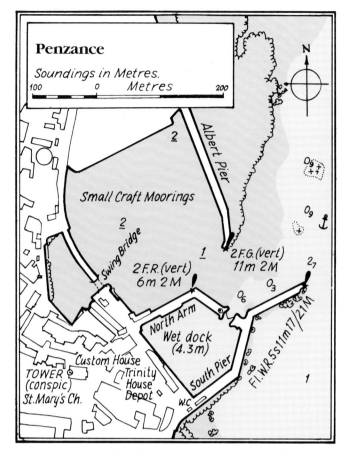

permanent collection of the Newlyn Painting School. They provide a romanticised glimpse of the final years as a simple working port, just before Brunel's Great Western Railway wiped out the isolation of the far west, and the first large hotels began to rise along the new promenade at the shingly head of the bay. There's nothing like a bit of nostalgia; conveniently we tend to forget all that coal dust.

PORT GUIDE: PENZANCE
AREA TELEPHONE CODE: 01736

Harbour Master: Captain Martin Tregoning, Harbour Office, North Arm, Penzance Harbour, Penzance, TR18 4AB. (Tel: 66113) Mon-Fri 0830 – 1200, 1300 – 1630
VHF: Channel 16, works 12, call sign *Penzance Harbour Radio* (office hours and HW-2 to HW+1)
Mail Drop: Harbour Office
Emergency Services: Lifeboat at Newlyn. Falmouth Coastguard
Anchorages: In fair weather 350m ENE of Albert Pier, clear of fairway or similar distance due south of South Pier
Mooring/berthing: Tidal wet dock (maintained depth 4.3m) entry 2 hours before to 1 hour after local HW. Anchor off or berth inside South Pier if Isles of Scilly ferry berth available
Dinghy landings: Steps on inside of Albert Pier
Marinas: None
Charges: 27p per foot inc VAT
Phones: Nearest public phone on promenade
Doctor: (Tel: 63340)
Hospital: West Cornwall Hospital (Tel: 62382)
Churches: All denominations
Local weather forecast: Harbour office
Fuel: Diesel by arrangement with Berthing master. Petrol in cans from garage
Gas: Calor/Gaz from Berthing master
Water: See Berthing master
Tourist information office: At railway station
Banks/cashpoints: All main banks, all with cashpoints
Post Office: Market Jew Street
Rubbish: Bins on quay
Showers/toilets: Under harbour office, coded entry lock
Launderette: Near railway station
Provisions: All normal shops, EC Weds but mostly open. Tesco in Market Jew Street. Large Safeway supermarket at Long Rock, bus/taxi ride away, open Sunday in season
Chandler: Matthews, New Street (Tel: 64004). Penzance Chandlers, Wharf Road (Tel: 67851) HM can arrange chandlery from Falmouth within 24 hours
Repairs: Drying out by arrangement with Harbour Master
Marine engineers: Albert Pier Engineering (Tel: 63566). R&D Engineering (Tel: 60253). Mounts Bay Engineering (Tel: 63095)
Electronic engineers: Kernow Marine Electronics, (Tel: 68606)
Sailmakers: Matthews, New Street (Tel: 64004)
Transport: Main line rail. Buses. Ferry and helicopter service to Isles of Scilly
Car Hire: 'Budget' at railway station (Tel: 330036)
Taxi: (Tel: 50666)
Car Parking: Large car park close to harbour
Yacht Clubs: Penzance Sailing Club, Albert Pier (Tel: 64989)
Eating out: Good choice pubs to restaurants
Things to do: Trinity House Lighthouse Centre. Nautical museum. Penzance Town museum. Trips to Lands End/ St Michaels Mount

largest town in the far west and just about all normal requirements will be found. The main shopping centre, the curiously named Market Jew Street derived its name from the Cornish 'Marghas de Yow' meaning 'Thursday Market' and climbs up the hillside past the main Post Office towards the impressive granite Market House, with its Ionic columns and domed roof, which also houses Lloyds Bank. In front of it there is a statue of Penzance's famous son, Sir Humphrey Davy, best known for his invention of the miner's safety lamp. Close to the station, on the road out of town, there is a good launderette open seven days a week, 0900-1930. There is a wide choice of options for eating ashore from Indian to fish and chips, with several good pubs in the area, notably the Turks Head, Dolphin and the Admiral Benbow. Chapel Street, leading back down from the top of Market Jew Street towards St Mary's church and the harbour, has a number of fine listed buildings, many of them gift shops, and also the bizarre Egyptian House, restored by the Landmark Trust, and the Nautical Museum, a private collection of fascinating artefacts salvaged from local wrecks, open daily May to September.

If time permits, a trip inland to the north coast or Land's End is well worth the effort, and there are regular bus services to St Michael's Mount if you decide against a visit by sea. Should the weather scupper your plans to sail to the Isles of Scilly, you could always console yourself with a day trip to the islands. As well as the *Scillonian's* daily sailings there are also frequent helicopter flights from nearby Penzance Heliport.

The small town Museum is located centrally in the Penlee Memorial Park. Apart from history and old photographs of the harbour, it is also the home of the town's

NEWLYN

Tides: HW Dover +0550. Range: MHWS 5.6m – MHWN 4.4m. MLWN 2.0m – MLWS 0.7m.
               Main harbour dredged to 2.4m LAT
Charts: BA: 2345. Stanford: 13. Imray: C7
Waypoint: South Pier Head 50°06'15N. 05°32'50W
Hazards: Low Lee Rock (lit). Gear Rock (unlit). Busy fishing harbour, beware vessels in narrow entrance. Approach dangerous in strong southerly weather, heavy swell sets across entrance. Pot and net buoys in Mount's Bay

Walk the streets of Newlyn at your peril! Large articulated refrigerated lorries, fork lift trucks and spraying hoses create an assault on all sides, leaving little doubt that this is a very busy fishing port, alive with colour, boats and activity. In contrast with its declining neighbour Penzance, Newlyn saw considerable commercial expansion during the 1980s after the opening of the 'Mary Williams' pier, and is now the largest fishmarket in England.

The port is dredged to an average depth of 2.4m LAT, and is accessible at all states of the tide, providing the only real harbour of refuge in Mount's Bay. However, in strong south or south-easterly winds a heavy sea builds up in the shoaling water at the head of the bay, particularly at LW causing a considerable run across the harbour entrance. This should be carefully considered if running for shelter, and entry is best attempted as close to HW as possible.

Approaches

Approaching from the east the run across Mount's Bay is straightforward, passing to the south of the unlit 'Mountamopus' YB south cardinal buoy and well clear of Gear Rock, three-quarters of a mile to the north-east of the harbour entrance, which dries 1.8m and is marked by an isolated danger beacon. From the west, Low Lee shoal, least depth 1.5m LAT, is marked by the 'Low Lee' BY east cardinal buoy (Q (3)10s), but Carn Base shoal three cables to the north-west is unmarked, has a least depth of 1.8m, and breaks in heavy weather. However, once St Clement's

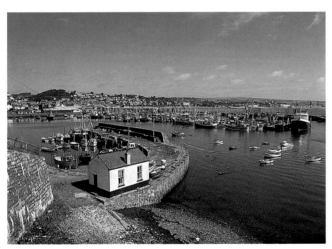

Yachts lie alongside the locals in this busy fishing port

Island off Mousehole is abeam, a course should be held just over a cable from the shore which will lead inside all the shoals, past Penlee Point with its former lifeboat house from where the white lighthouse on the end of the south pier (Fl 5s visible 253°-336°T) is easy to spot against the town which rises up the hillside overlooking the harbour. The entrance is 47m wide and the northern side has a (FWG) sectored light, green showing 238°-248°T. The only other hazard to consider, particularly at night, is the considerable intensity of pots along this stretch of coast, and also the large numbers of small craft that will be found working them; as always, fishing craft restricted by gear should be given a generous berth, and your intentions made obvious at an early stage.

Anchorage and berthing

In offshore winds it is possible to anchor 200m SSE of the harbour entrance, sounding in to about 3m, or alternatively a similar distance to the north-east in Gwavas Lake. Owing to the frequency of fishing boat movements, a riding light is essential. Newlyn provides no specific facilities for yachts, but overnight visitors will usually find space alongside the fishing boats on the west side of the central 'Mary Williams' pier, which tends to be occupied by local vessels that are laid up for refits or repair. There is less movement here in contrast to the continual comings and goings elsewhere in the harbour where the fleet lands daily. *As in all busy fishing harbours remember that commercial activity takes precedence over pleasure. Sailing and anchoring is prohibited anywhere in the harbour except in an emergency.*

As well as breast lines to your neighbour, bow and stern lines should be taken ashore if intending to stay overnight, and with the boats often five or six deep this will involve quite a scramble. The best bet is to moor temporarily, and check with the Harbour Master, Mr Munson, whose office is at the head of the harbour by the north pier, above the fish-merchants' offices. He is very helpful, and will endeavour to find you a quiet berth. However, yachts on passage, or sheltering from weather are only allowed to stay on a strictly day to day basis. It is essential to let the Harbour Master know if the entire crew is going ashore for any length of time and always best, to leave someone on board. The wet dock in Penzance is the nearest place where a yacht can be left unattended for any length of time.

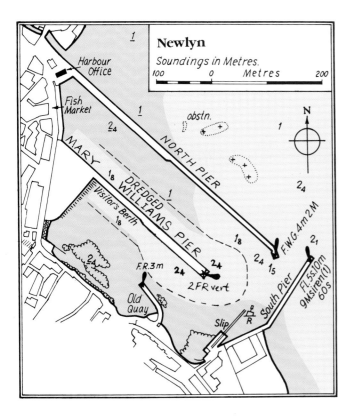

Facilities

Newlyn is a bustling working town, the fishmarket central to its life, and early risers, providing they keep out of the way of the very serious business in hand, can spend a fascinating hour or two watching the landings and daily auction which starts at around seven in the morning, Monday to Saturday. The old town, a pleasant meandering sort of place, climbs up the hillside in tight rows of sturdy granite cottages, sheltered courtyards and narrow alleyways. It has been a fishing harbour since medieval times, and the magnificent remains of the original pier, a gently curving wall of huge granite blocks, weathered and mottled with orange lichens can still be seen forming part of the small inner basin beneath the cliff on the west side of the harbour.

The present harbour, built between 1866 and 1888, was the centre of the huge mackerel and pilchard fishery, and in 1896 the scene of the infamous riots when the devout local fishermen ran amok, angered by visiting east coast boats landing fish on Sundays. Hurling their catches back into the sea, the bloody disturbance that ensued was only eventually put down by military intervention. Renowned for their speed and sea-keeping ability, the magnificent 30-50 foot luggers that once filled these west Cornish ports have mostly vanished, but fortunately *Barnabas*, a fine example of a St Ives lugger restored and owned by the Maritime Trust, is based in Falmouth where she can often be seen sailing during the summer months.

The present fleet has grown considerably in recent years, and ranges from large beam trawlers and scallopers that venture as far afield as the North Sea and the Irish Sea to the smaller pot and net boats that work the tricky inshore waters round Land's End and fish for mackerel in winter.

Although small, the town provides most necessary provisions, catering as it does for the fishing fleet, and

supplies can be obtained on Sundays from 'Caddys' close to the harbour, as most of the fleet puts to sea on Sunday evening. Anything unobtainable in Newlyn can usually be found in Penzance, a pleasant walk along the promenade, or by bus every fifteen minutes. Hot showers, for both men and women, can be obtained at the 'Royal National Mission to Deep Sea Fishermen'.

This part of Cornwall has a long tradition of popularity with artists, in particular the famous Newlyn School of the 1890s, whose realistic paintings have enjoyed a great resurgence of interest, with fishing and fisherfolk one of its central themes. Today's cultural diversion is provided by the Newlyn Art Gallery on the seafront with regular local and visiting exhibitions; gastronomic diversion is well catered for, with several fishermen's cafes, fish and chips (very fresh!), a Chinese takeaway, some good pubs with food, notably the Tolcarne, and the Smuggler's which overlooks the harbour with an adjoining restaurant. There are also several more up-market bistros, with, predictably, a good choice of seafood.

PORT GUIDE: NEWLYN
AREA TELEPHONE CODE: 01736

Harbour Master: Mr Andrew Munson, Harbour Office, Newlyn (Tel: 62523) Mon-Fri 0800-1700
VHF: Channel 16, works 12 and 9, call sign *Newlyn Harbour* office hours only
Mail Drop: c/o Harbour office
Emergency Services: Lifeboat at Newlyn. Falmouth Coastguard
Anchorages: In fair weather to NE or SE of harbour entrance clear of approaches. No anchoring within harbour
Mooring/berthing: Alongside fishing boats on west side of central Mary Willams pier
Charges: Overnight for 30 footer £6.00 inc VAT. 50% surcharge for multihulls
Phones: By Harbour office
Doctor: (Tel: 63340)
Hospital: (Tel: 62382)
Churches: All denominations
Local weather forecast: At Harbour office
Fuel: See Harbour office for diesel. Petrol in cans from garage
Gas: See Harbour office
Water: Available on quay
Banks/cashpoints: Barclays and Lloyds, Mon-Fri 1000-1230 no cashpoints. All banks in Penzance have cashpoints
Post Office: By Harbour office
Rubbish: Bins on quays
Showers/toilets: Royal National Mission to Deep Sea Fishermen, North pier. Public toilets on Quay
Launderette: On road towards Penzance
Provisions: Supermarket, butcher, newsagent. 'Caddys' stores open Sunday morning
Chandler: 'Cosalt', Harbour Road (Tel: 63094). 'South West Nets', Harbour Road (Tel: 60254)
Repairs: Large slip for commercial craft. Drying out by arrangement with Harbour Master
Marine engineers: C K Jones, North pier (Tel: 63095). See also Penzance port guide
Electronic engineers: Kernow Marine Electronics (Tel: 68606). Sea Com Electronics (Tel: 69695)
Sailmakers: See Penzance port guide
Transport: Buses to Penzance. Main rail line at Penzance
Car Hire/taxi: See Harbour Master
Car Parking: Near Fishmarket
Eating out: Fish and chips, pub food, and bistros

ST MICHAEL'S MOUNT AND MOUSEHOLE

St Michael's Mount:	HW Dover +0550
Charts:	BA 2345. Stanford 13. Imray C7
Hazards:	Harbour dries. Hogus rocks, Outer Penzeath rocks and Maltman rock (all unlit)
Mousehole:	HW Dover +0550
Charts:	BA 2345. Stanford13. Imray C7
Hazards:	St Clement's Island (unlit). Harbour dries

Providing an interesting alternative to the larger Mount's Bay ports of Penzance and Newlyn, St Michael's Mount and the old fishing harbour of Mousehole are well worth a visit if conditions permit. Both have small harbours which dry completely at LAT but are normally accessible after half flood.

St Michael's Mount

There is no problem identifying St Michael's Mount, Cornwall's own mini version of the famous Mont St Michel in Normandy – this dramatic tidal island rises into a distinctive 90m pyramid in the north-east of Mount's Bay, with a fairy tale castle at its summit, in which the St Aubyn family have lived since the mid-1600s. They continue to do so, but the island was presented to the National Trust in 1954 by the present Lord St Levan. During the summer months, the Mount and parts of the castle buildings are open to the public on weekdays, 1000-1700; there is a charge for admission.

Historically, the tiny harbour on the northern side of the Mount was once the most important in Mount's Bay, a major centre for the export of tin chronicled by the Romans in the first century BC, when the ore was sensibly brought overland from the mines on the north coast to avoid the treacherous journey around Land's End. A Benedictine monastery was established here in 1135, making it an important place of pilgrimage, but this was dissolved by Henry VIII during the Reformation, and during the Civil War the Royalists held the Mount for four years, a vital siege as the fortress was an ideal place for the import and stockpiling of arms and ammunition brought in from France.

In 1727 the modest harbour was rebuilt and extended, and by the early 1800s it had prospered considerably, with over fifty houses on the island and a population of about 300. As you will see from the many pictures and prints if you visit the castle, large numbers of vessels regularly lay here, anchored in the bay or crammed into the tiny harbour, discharging timber from Scandinavia, coal and salt, before loading the return cargoes of copper, tin and cured fish that was exported in considerable quantities all over Europe. It was, however, the new harbour at Penzance that eventually sealed the fate of the Mount, and by the turn of the century it had lost all the trade to its larger rival.

Approaches and anchorage

The best anchorage for a visit to the Mount is just to the north-west of the harbour entrance; alternatively it is possible to lie alongside in the harbour where you will dry out on hard sand. The approaches have a number of rocks and shoals and should only be considered in fine weather, offshore winds, and a rising tide. Do not attempt it at night, and under no circumstances should an approach ever be made to the east and north of the Mount where there are extensive rocky shoals.

From **Penzance**, keep offshore to avoid the reefs across the head of the bay; Western Cressar Rocks and Ryeman Rocks are both marked by a YB south cardinal beacon and a course towards the southern extremity of the Mount will clear both these and the Outer Penzeath rock, awash LAT and unmarked. Closing the Mount, approach on a north-easterly course keeping a good 300m from its steep western side to avoid Guthen Rocks, a shoal patch with just over 2m LAT, due west of the castle. Sound in towards the anchorage where about 2m will be found at LW, over a firm sandy bottom between the Great Hogus reef, which dries 4.9m and the western pier end.

After half-tide, the harbour is accessible to average draught boats – stay a reasonable distance from the pier

Visitors dry out in the small harbour on St Michael's Mount or anchor between it and the Great Hogus Rocks

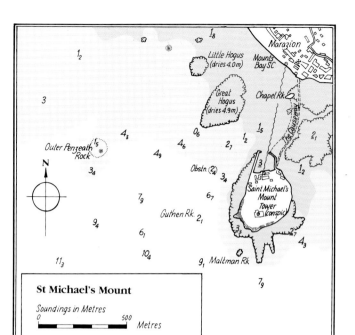

St Michael's Mount
Soundings in Metres
0   500
Metres

of a Cornish fishing village, though long past its heyday during the mackerel and pilchard fishery. The residue of the fishing fleet once based here is now kept in Newlyn but a few local pot, net, and angling boats lie here on fore and aft trots and there are a number of small pleasure craft, too.

From early November until the end of March the harbour is closed completely with heavy wood baulks across the mouth and few winters pass without some damage to the seemingly impregnable walls as the south-easterly gales roll into Mount's Bay. Privately administered by its own Harbour Commissioners, the unfortunate dominance of car parking beside the harbour and on the quays does pay for its upkeep, and the small charge for visiting yachts contributes to it too. In suitably settled conditions, the harbour, or the anchorage off it provide a worthwhile diversion for an overnight stay, or just a daytime visit.

Approaches and anchorage

Approaching across Mounts Bay the houses of Mousehole climbing the hillside are easy to spot to the south of Newlyn, although St Clement's Isle, a low rocky outcrop with a small obelisk on its highest point just east of the harbour mouth, will be lost against the land until much closer. Approaching from the south, from Penzer Point the island is much easier to see as it lies clear of the land. Between St Clement's Island and the harbour entrance there is a convenient anchorage, rolly at times, but particularly handy if waiting for sufficient water to enter.

Although local fishing boats use the inshore passage to the north of the island there are rocks on both sides and a very narrow channel at LW. Strangers should always use the approach from the south of the island which is far wider and safer, and enter mid-way between the island and the shore. Beware Tom Kneebone ledge, least depth 0.9m LAT, 100m south of the obelisk, do not cut the corner but enter on a north-westerly course, sounding in to a point midway between the middle of the island and the south pier head where a depth of 3m will be found at LW. An anchor light is advisable. Closer inshore, about 100m south and slightly east of the entrance, isolated Chimney rock dries 2.3m LAT but the biggest hazard is probably the many pot buoys around the island and along the shore – be particularly wary in the approach.

Berthing

After half-tide there is plenty of water inside the harbour; on the outer corner of the northern wall there are a number of horizontal concrete ledges and if only the top three are showing there will be a good eight feet inside the outer ends of both harbour walls where you can dry alongside on a firm sandy bottom. The northern arm tends to be the better spot, with three convenient ladders. On both piers there are stone steps which should not be obstructed as these are used by the local boats for landing fish and crews. Alternatively, bilge-keelers can anchor fore and aft if space can be found among the local boats where they will dry out. However, the possibility of fouling the moorings does make the quay the better option providing you don't mind becoming an inevitable object of constant fascination to the passing holidaymakers. There is no harbour

heads and beware of the constant stream of ferry boats. Known locally as the 'hobblers' they operate from the steps on both sides of the entrance continuously as soon as the tide covers the causeway and should not be obstructed in any way. Berth between the ladders on the western wall and report to Len Ritchie, the Harbour Master whose office is the opposite side of the harbour; if you can't find him immediately don't worry – he is also the island's Postman and is probably dealing with the mail! A charge of about £7.50 will be made for a night's stay, well worth it for the eventual peace and quiet.

Facilities

Today, the Mount is the focal point of a very different kind of industry – tourism – and at times is almost overrun by the hordes of holidaymakers streaming like the Israelites across the Low Water causeway from Marazion with a wary eye on the tide, or packing the continuous fleet of small ferries that run during the summer months when the tide is in. In spite of it, the small harbour, with its simple row of neatly restored cottages, is well worth a visit. Despite the dire warning that the climb to the castle 'should not be attempted by those suffering from any kind of heart condition', it is not as steep as it looks and provides spectacular panoramic views across Mount's Bay. In the evening, when the crowds have gone this is once again, for a few brief hours, a peaceful and tranquil place. There is a restaurant, cafe and gift shop by the harbour when the Mount is open and the Harbour Master can fill a container of water if you're desperate. Most normal provisions, Barclays and Lloyds bank, phone and Post Office and several pubs will be found at Marazion.

MOUSEHOLE

Mousehole is in many ways similar to St Michael's Mount, for this is another picturesque honey pot around which the tourists swarm. Available from half tide, this small, oval harbour with massive granite boulder walls is the epitome

121

office, and the Harbour Master, Frank Wallis, will probably wander down in the evening to collect his dues, about £3 a night, with a surcharge for multihulls.

A bustling, colourful place where visitors bravely compete with the traffic trying to squeeze through the narrow streets, Mousehole's unique charm has not only survived the onslaught of recent years but also considerable mayhem in the Middle Ages when the village, then the most important port in Cornwall, was raided and burnt to the ground by marauding Spaniards. It enjoys a certain fame today as the home of Dolly Pentreath who died in 1777, reputedly the last person to speak Cornish as her native tongue, and a commemorative plaque can be found on the wall of her cottage close by the harbour. Perhaps she would have been able to provide the answer (although presumably few of us would have been able to understand it . . .) to the village's curious name for which there seems to be no definitive explanation. Remember, though, that it is always pronounced as 'Mowzull', and never, *never* as Mouse Hole!

After Dolly Pentreath, Tom Bawcock is probably the other most celebrated former inhabitant, and 'Tom Bawcock's Eve' on 23rd December recalls the time when this fabled local fisherman put to sea in desperation after weeks of gales, returning with a fine catch of seven different kinds of fish to provide Christmas feast for the starving village. Hopefully, you will not be visiting by sea at this time of year – not only will you find the baulks down, but you will also be confronted by the infamous 'Starry Gazy Pie', a hideous creation with the heads of the seven varieties of fish poking through its crust.

Facilities

Starry Gazy apart, most other basic provisions are available in the village: groceries, off-licence, Post Office, butcher and newsagent. Just about everything else is obtainable in Penzance and there is a regular bus service; for the more energetic, bikes can be hired just outside the village on the Newlyn road. There are several cafes, the Ship Inn, in spite of the tourism has managed to retain a distinctively village pub atmosphere, and though present throughout the village, the tragedy of the Penlee lifeboat particularly

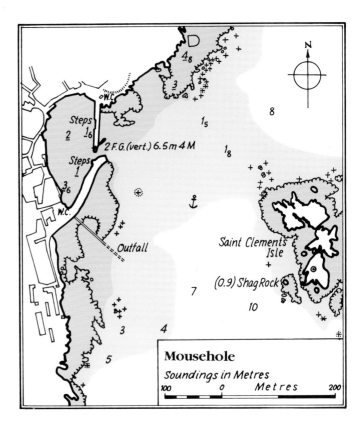

lingers here as the former landlord of the pub was one of her crew. There is a small, reasonably priced restaurant adjoining the Ship but for those with more expensive tastes the 'Lobster Pot' overlooking the harbour has a worldwide reputation. Whether you need to work up an appetite or not, the fine walk along the cliffs to the south of the village towards Penzer Point and the beautifully named Lamorna Cove should not be missed.

Mousehole, like many of its counterparts in Cornwall has a famous male voice choir. If you are lucky, on one of those increasingly rare, balmy summer's evenings, sometimes they assemble on the quayside, soft Cornish voices drifting in a strangely haunting harmony across the natural arena of the harbour into the gathering twilight around the bay. It will be one of those particularly evocative moments that you, like me, will never forget.

Mousehole at LW. St Clements in the distance

THE ISLES OF SCILLY

Tides: HW Dover +0607 Range: St Mary's MHWS 5.7m – MHWN 4.3m MLWN 2.0m – MLWS 0.7m.
        See special note on tides, page 126
Charts: BA: 34, 883, 1148 2565. Stanford: 2 Imray: C7
Waypoints: See *Passages, Mainland to Scilly*
Hazards: Many unmarked rocks and shallow ledges, large areas between islands dry. No harbour offering all weather security

'A straggling collection of barren rocks, wonderfully broken up, narrow strips of land bordered with marvellously white gleaming beaches, rocky hills covered with heath and gorse, intersected by ravines running down to the loveliest little bays, piles and piles of strangely weird grey, lichen covered rocks, little dales where bask a few isolated cottages festooned with mesembryantheums and giant geraniums sheltered by a few tangled tamarisks . . .' Frank Cowper, on his first visit to the Isles of Scilly in 1892 was clearly impressed by what he had found; over a century on, his appraisal still holds good.

Set in water of breathtaking clarity this tantalising archipelago of 48 islands, some little more than glorified rocks, extends over an area of approximately 45 square miles between 21 and 31 miles WSW of Lands End. Only six islands are inhabited, St Mary's, St Martin's, Tresco, Bryher, St Agnes and Gugh. Of the rest, eighteen are described by the Admiralty Channel Pilot as *'being capable of bearing grass, the remainder are barren'*.

The warmer Gulf Stream climate prompted the Romans to name the area *Sillinae Insulae*, meaning Sun Isles, from which Scilly now derives its name, in itself, something of a moot point. Although many loosely refer to the islands as the 'Scillies', (and I have been guilty of this), the locals, let it be known, prefer *Isles of Scilly* or just plain *Scilly*. Nor do they take too happily to being described as Cornish; they are *Scillonians*, most definitely, and proud of it . . .

In fine and settled weather it is easy to imagine, if

'White gleaming beaches, rocky hills covered with gorse and heather . . .' St Helen's Pool looking towards Tresco

123

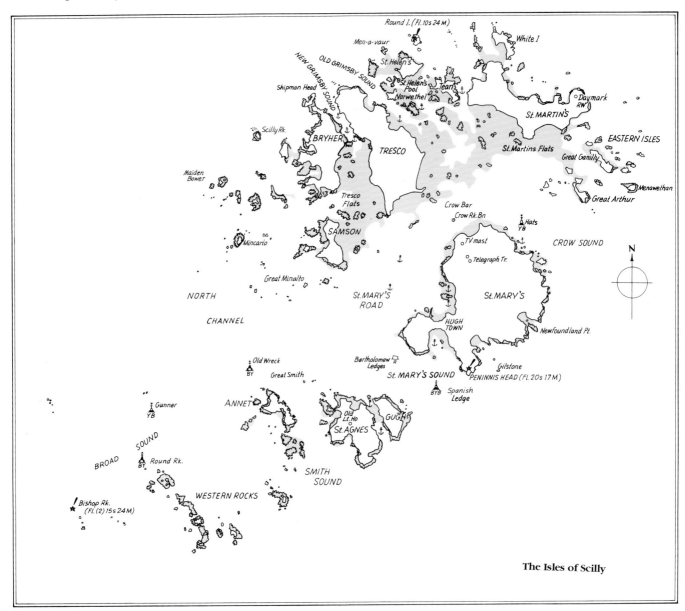

The Isles of Scilly

momentarily, that you are in a far more exotic location, given a few more palm trees in addition to those that already flourish here. Certainly, it is often hard to believe that you are still so close to mainland Britain.

However, such delights are, alas, not always so easily enjoyed! As a cruising ground the Isles of Scilly can have a definite downside. It is not an area to be trifled with, and a clear understanding of the potential problems will go a long way to ensuring a successful visit.

The first consideration is getting there. It will involve an offshore passage of over 20 miles, admittedly not a great distance, but one that takes you out into the Atlantic and a particularly exposed sea area renowned for sudden changes of weather and poor visibility. With strong tides, shipping, including the Land's End Traffic Separation Scheme (TSS), and the rarely absent Atlantic ground swell this is no place for the inexperienced. Even for the experienced, it can be a very unpleasant stretch of water to get caught out in.

The second consideration is Scilly itself. Apart from the obvious dangers of any group of islands strewn with rocks, most unmarked, large areas of shallow water and strong

and often unpredictable localised tidal streams and races, it also lacks a harbour or anchorage that can be considered secure in all weather. As each anchorage is only safe in certain combinations of wind and sea any amount of time spent in the islands will inevitably involve shifting as the weather changes, often in deteriorating conditions.

Once among the islands, distances are not great, usually only a matter of a few miles, but the problems of riding out bad weather have been greatly exacerbated by the demands made on the available anchorages by the ever increasing numbers of visiting yachts, particularly during August when large numbers of French boats abound.

The third consideration is the relative lack of facilities for visiting yachts when compared with those now taken for granted on the mainland. With the exception of **New Grimsby Sound** where there is a limited number of visitors' moorings, in most other places you will not be able to lie to a mooring or get alongside a convenient quay or pontoon, *and you will have to anchor.* Simple enough, one might think, but the holding ground in most places is of fairly indifferent quality, either fine loose sand, sand covered with weed or weed covered rocks.

Finally, owners of deeper draught boats should be aware of one other little mentioned factor which can at times seriously detract from full enjoyment of Scilly – the all-pervading Atlantic ground swell. This was particularly evident during July 1994 when much of this research was carried out and it plagued us in most anchorages, usually at its worst towards HW and when the wind was at its lightest – calm weather does not necessarily mean a calm sea in Scilly.

As one restless night succeeded another I became convinced that the ideal craft for exploration of these waters is a boat of moderate draught, able to edge close inshore or, even better, a bilge keeler able to dry out completely, for then the options for a more relaxing stay increase considerably.

However, the object of all this is not to deter, although this is often the reaction on first perusal of the large scale chart. The hazards seem myriad and the pilotage overly complicated, not least the abundance of **transits** which are traditionally associated with Scilly as leading and clearing lines for passages and hazards. For many, these can prove overly confusing, particularly on first arrival when the topography is itself bewildering enough. The 307°T leading line to clear the Spanish and Bartholomew Ledges in St Mary's sound, *'North Carn on Mincarlo in line with south side of Great Minalto'*, is a typical example, as Great Minalto, in spite of its name, is a actually a much smaller rock than Mincarlo. . . .

The leading lines are not critical to navigation and only an aid as long as you can identify them – often the visibility is such that the more distant marks are indistinct, if not invisible.

Transits are also useful to gauge any tidal set across your course, but if you do decide to try using them, leave nothing to chance. Many rocks have a tendency to look similar – once you believe you have identified your leading marks, check that they tally with the heading on the compass (remembering that all bearings are True from seaward and will need to be corrected for magnetic). If in any doubt, don't panic, concentrate instead on making your approach in the normal fashion using any other visible criteria you have at your disposal.

The choice of anchorages: I have limited to those most likely to be used during a first visit, and I am also assuming that most readers will be approaching from the mainland. The western side of the islands has by far the greater number of isolated offshore hazards and in my opinion it should be avoided, certainly by newcomers to the area. For this reason the western approaches through **Broad Sound**, **Smith Sound** and the **North Channel** are deliberately not covered in any great detail. If you are keen to see the Western Rocks at close quarters there is much to recommend a trip in one of the local pleasure boats . . .

There are inevitably numerous other possible anchorages and more intricate passages requiring suitable weather and local knowledge that might appeal to the more adventurous, or those who have already explored the more familiar places. However, my only intention here is to try and ensure a safe and pleasant stay in these often challenging waters. To this end, my own pointers for a first

and, hopefully, trouble free visit to the islands would be summarised as follows:

1) Have a well found boat with a reliable engine, a crew of adequate strength and experience, and, absolutely essential, up to date editions of Admiralty charts 883 (Isles of Scilly, St Mary's and the principle Off-islands) and 34 (Isles of Scilly) which includes tidal charts for the islands. *The chartlets accompanying this text are simplified and much detail has been omitted to assist clarity. They must not be used for navigation.*

Familiarise yourself with the layout of the area in advance. Although the pilotage seems complicated it is, for the most part, more straightforward after you have got your initial bearings. Once you have arrived make every effort to identify and memorise salient features – it is surprising how quickly you begin to feel more at home.

The clarity of the water enables the rare treat of eyeball navigation and a good lookout will be able to spot most dangers long before you hit them. Most of the rocks are covered with long growths of bright green weed that is usually visible on or near the surface at anything other than HW.

2) Ideally, choose a spell of fine settled weather, with Neap tides, and arrive in daylight with ample time in hand to find a suitable berth. Do not be tempted to make a dash for the islands if there is any hint of deteriorating weather in the offing. You are far better off remaining on the mainland.

3) Try to visit the islands in June or July. If possible avoid August when they are at their busiest and the weather has a tendency to be less settled.

4) Have more than an adequate amount of ground tackle on board to cope with all eventualities. Most boats will normally have a CQR, Bruce or Danforth as their main anchor but here you cannot afford to skimp on the size of your kedge(s). With the poor holding and fine sand in many anchorages I have found that my CQR does not always perform at its normal best and most local boatmen seem to favour a good sized Fisherman or Danforth. I would recommend one of each in addition to your normal ground tackle, with ample chain and warp of adequate weight.

Be sure you know how to use it all. Lower anchor until you feel it touch the ground, go astern and slowly veer chain to lay it out along the bottom, allowing at least 3 times the anticipated depth at HW (5 times with warp). Belay cable, and give boat a good long burst astern to ensure that the anchor bites. If the chain or warp jumps and loses tension it has definitely not taken; don't take a chance, but start again.

There is much sense in the old maxim that it is better to be safe than sorry and always lay a kedge. Certainly, *if you are leaving your vessel unattended for any length of time this is definitely recommended.*

5) Keep a careful watch on the weather and be constantly aware of any potential changes, particularly the direction in which the wind is most likely to shift. Accordingly, *choose your anchorage not only with regard to the prevailing conditions but also what might happen within the next 12 hours.* If remaining anywhere overnight be certain that you will be able to leave in the dark if necessary (work out a course to steer in daylight) and at all states of the tide. Always have an alternative anchorage in mind before the you need to find one.

In addition to the normal BBC shipping forecasts, Radio Cornwall shipping and inshore forecasts can be obtained in Scilly on FM 96.0. Land's End Radio transmits shipping forecasts and traffic lists to Scilly on VHF channel 64 at 0733 and 1933 UT.

If there is a serious likelihood of bad weather in the offing and the timing permits, do not overlook the possibility of making a speedy return to the mainland before its onset. Remember, though, that Mounts Bay can be dangerous to approach in heavy weather from the south-east and south and that the tidal dock at Penzance is only accessible from 2 hours before to 1 hour after local HW.

6) Once in the islands don't hurry, take life at a gentler pace and don't take chances, least of all with the tide. The tidal streams are very unpredictable in the close proximity to the islands and for short periods they can often attain much greater strength than indicated in the Admiralty chartlets and Pilot. Predicted tidal heights are particularly susceptible to the atmospheric pressure – during prolonged spells of high pressure they can sometimes be almost a metre less than anticipated, and remember, too, that the 5m range at Springs will considerably reduce the amount of space in many anchorages, conversely, with a 2.3m range at Neaps there is a lot more water available.

However, if you plan your moves on a rising tide, ideally after half flood, most of the trickier, shallower passages can be tackled with impunity. Try them once the ebb has set in and you've no one to blame except yourself when things go wrong!

Following the local trip boats is *not* recommended, their Skippers know these waters intimately, many of their short cuts pass rocks within a hair's breadth, and they seem able to skim over the shallows on little more than a heavy dew. Be particularly wary of the large Bryher ferry, *Firethorn;* in spite of her size, she draws little more than 2 feet!

7) Do not impede local boatmen, fishermen or other commercial craft; they are earning a living, you are there for fun, and they do not take kindly to finding yachts lying on their moorings when they return after a hard day's work, or blocking the quay steps when trying to embark or disembark passengers. If you are in any doubt – ask!

8) Water is a valuable commodity in Scilly, and you will have to pay to obtain it alongside. Make sure you fill your tanks on the mainland. As food and booze (which is all freighted in by sea or air) is inevitably more expensive it is also worth stocking up before you leave. Gas refills are only obtainable on St Mary's so make sure you always have a spare.

9) A number of the uninhabited islands are important bird breeding areas. The following are closed to visitors between 15th April and 20th August: Annet, The Western Rocks, Crebewethan, Gorregan, Melledgan, Rosevear, Norrard Rocks, Castle Bryher, Illiswillgig, Maiden Bower, Mincarlo, Scilly Rock, Stony Island and Green Island (off Samson) and Men-a-vaur.

In addition Tean is closed on a voluntary basis between 15th April and 20th July when ringed plovers and terns are nesting, and certain areas of other islands have clearly marked nesting sites, such as the southern end of Gugh which visitors are asked to avoid. Nesting and territorial birds, particularly gulls, can be extremely aggressive. They will attempt to ward off intruders with alarming swoops to the head that are potentially injurious and definitely frightening for smaller children and timid souls like me.

10) Beware pot buoys! They will be encountered both in the outer approaches and anywhere among the islands. With such a wide selection to choose from wreck diving is also very popular, and you should give any boat flying international code flag 'A' (vertical white with blue swallow tail) a wide berth.

As Cowper concluded, *'A stay of a few days in these bewildering islands should afford most people a good deal of pleasure. There is perpetual variety. Every rock and bay and hill offers some new view, and the contemplation of this decomposing heap of stones in the midst of the ever-vexed Atlantic must arouse a wondering curiosity, if not an enthusiastic admiration . . .'*

Pilot gig, St Martin's. Tean Sound beyond

Passages
MAINLAND TO SCILLY

Given a well-found boat, suitable weather and preparation the passage from the mainland should pose no major problem. Given unsuitable weather it should not even be contemplated.

On a clear day, from the high ground of Land's End the islands are often clearly visible, but with the daymark on St Martin's rising to a mere 56m and the highest ground on St Mary's 48m, normally, if you're lucky, they do not begin to show from sea level until about 12 miles distant; more often than not, it will be much less. As you sail ever further away from the land and, it seems, far out into the Atlantic, the doubts do not take long to crowd in – it is always worth remembering that one tends to anticipate a landfall a lot sooner than it usually appears.

St Martin's daymark looking north-west to Round Island

The sight of the all-white ferry *RMS Scillonian III* helps to dispel a bit of unease – she sails a direct course from the Runnel Stone to St Mary's, leaving Penzance daily at 0915, and returning from St Mary's at 1630, except on Saturdays during late July and August, when she makes two trips, leaving Penzance 0630 and 1345, departing St Mary's 0945 and 1700. There are very occasionally, Sunday sailings. Depending on the tide and weather she makes her approach into the islands either through Crow Sound or St Mary's Sound.

The frequent 'British Airways' helicopters are direct flights from Penzance to St Mary's or Tresco; the Airbus flights to St Mary's are from Lands End Airport, on the north Cornish coast. In the season, the odds are also very much in favour of there being other pleasure craft bound in either direction . . .

Those armed with a battery of electronics will perhaps fare better mentally, radar will doubtless confirm the islands existence from afar and also the Racons on Wolf Rock (3 & 10cm 10M Morse 'T'), Bishop Rock (3 and 10cm 18M Morse 'T') and Seven Stones (3 & 10cm 15M Morse 'O'). As long as the Americans keep the satellites on line GPS will doubtless do its stuff but be warned, due to base line extension Decca readings in the approaches to the islands are likely to be suspect, as too is the powerful RDF beacon on Round Island in the closer approach to St Mary's where the signal crosses the land. Here the aero beacon on St Mary's can prove useful although this is only operational during the daytime, Monday to Saturday. As always, a conventional DR plot should be maintained at all times.

Marine Beacons	Range (NM)	Frequency (kHz)	Call Sign
Lizard	70	284.5	LZ
Round Island	150	298.5	RR
Aero beacons (daytime only Mon-Sat)			
Penzance Heliport	15	333.0	PH
St Mary's airport	15	321.0	STM

In the event of fog, a sudden reduction in visibility, or an unexpected deterioration in the weather the option of aborting the passage and returning to the mainland should never be overlooked, however frustrating it might seem. In any of these conditions an approach to Scilly is risky at the very least, and potentially extremely dangerous. With so many offlying rocks and unpredictable tidal streams the fog signals should be regarded as a warning to stay well clear of the islands and not an invitation to attempt an approach: Round Island (Horn (4) 60s), Wolf Rock (Horn 30s), Longships (Horn 10s), Seven Stones (Horn (3) 60s) (do not confuse with Round Island), Bishop Rock (Horn Mo (N) 90s), Spanish Ledge buoy (bell) and Runnel Stone (bell and whistle). Tater Du (Horn 2 30s).

Do not be tempted to run for the islands in heavy weather in anticipation of finding shelter, the overfalls and heavy seas in the approaches during gale force conditions, particularly with wind against tide, could in themselves easily overwhelm small craft. Should you find youself in trouble all rescues in this area are co-ordinated by Falmouth MRCC VHF Channel 16 (works 67), call sign *Falmouth Coastguard.* There are lifeboats based at Newlyn, Sennen, St Ives and St Mary's.

A fine spell of settled anticyclonic weather would seem to be the ideal, for then the odds are in favour of a good easterly breeze to whisk you out to the west. However, in such conditions fine weather haze is more than likely to be prevailing and the visibility often much diminished.

Given the low lying nature of the islands there is much to be said for a night passage, timed to arrive soon after daybreak, for this busy stretch of water is well lit with the south cardinal 'Runnel Stone' buoy (Q6 + L Fl 15s) and the powerful lighthouses at Tater Du (Fl (3) 15s 23M), Wolf Rock (Fl 15s 23M), Longships (Iso WR 10s W19M R18/15M), Seven Stones Light float (Fl (3) 30s 25M), Round Island (Fl 10s 24M), and Bishop Rock (Fl (2) 15s 24M) all of which should be visible long before you would normally pick up any detail in daylight.

Although a night entry into the islands is not recommended for a first time visit it is feasible with extreme care using St Mary's Sound which is lit by Peninnis lighthouse (Fl 20s 17M), 'Bartholomew Ledge' buoy (Fl R 5s) and the leading marks into St Mary's Pool (FR over FR).

PASSAGE FROM THE SOUTH CORNISH COAST
Waypoints:
'Manacles' east cardinal buoy: 50°02´77N 05°01´85W.
Turning point 3M due S Lizard Light:
49°54´80N 05°11´90W
'Runnel Stone' south cardinal buoy:
50°01´15N 05°40´30W
Wolf Rock lighthouse: 49°56´70N 05°48´50W
'Hats' south cardinal buoy (Crow Sound):
49°56´17N 06°17´08W
'Spanish Ledge' east cardinal buoy (St Mary's Sound):
49°53´90N 06°18´80W

Bound for Scilly from the south Cornish coast you will most probably be making the passage direct from Falmouth or the Helford, a distance of just under 60 miles to St Mary's, or from Mounts Bay which reduces this to about 36 miles, Newlyn to St Mary's. In both cases the timing will invariably revolve around the tidal considerations, and the need to arrive in the islands in daylight.

Falmouth Bay to St Mary's.
To carry a fair tide south and west from Falmouth Bay around the tidal gate of the Lizard it is best to leave at about 3 hours after HW Falmouth (3 hours before HW Dover). This ensures a fair tide for the next six hours which should get you well across Mounts Bay and hopefully clear of the mainland. (See: 'Passages, Manacles to Land's End').

The rhumb line course of 270°T from a point about three miles south of the Lizard to Peninnis Head at the entrance to Saint Mary's Sound is a distance of 43 miles and will take you 2.5 miles south of Wolf Rock, which lies sevenmiles south-west of the Runnel Stone, providing a useful check on the tidal set.

Wolf Rock is always a useful check on your progress

Wolf Rock lighthouse is a slender grey granite column with a helipad and it can often prove difficult to spot at first, particularly if it has the morning sun upon it. Give the lighthouse a berth of at least ½ mile; although the rock only covers a small area it creates strong tidal eddies, and heavy overfalls immediately to the west in bad weather. Traditionally, the Wolf earned its curious name from its voracious appetite for passing ships and the first attempt to mark it in 1791 was a droll reflection of this. A 20 foot iron mast was erected on the rock by Lieutenant Henry Smith topped with a replica of wolf's head, its open jaws forming a sound box that produced a hideous wail as the wind blew through it. As predicted, it vanished in the first winter storms and it was not until 1840 that a 46 foot high stone beacon was erected by the Trinity House engineer James Walker, who later masterminded the existing lighthouse, a magnificent feat of engineering standing 110 feet above MHW.

Although begun in 1862 using pre-formed interlocking granite blocks shipped from Penzance, the structure was not completed until July 1869 and brought into service on 1st January 1870. Always regarded as one of the most difficult lights to relieve, it became the first offshore light to be fitted with a helipad in 1973. In common with most offshore lighthouses, it is now unmanned.

West of the Wolf you will find large vessels emerging from the **Land's End Traffic Separation Scheme** (TSS) The northbound traffic lane begins four miles west of the Longships, the southbound lane half-a-mile east of the

Seven Stones light float. Both lanes are three miles wide with a two mile wide separation zone and are marked on Admiralty charts 1148 and 2565.

As with all TSS you must cross the traffic lane *on a heading as near as practicable at right angles to the general direction of traffic flow.* This means that you must keep your vessel in full profile to the oncoming shipping with your fore and aft line at right angles to the traffic and not your track through the water. You must also cross the TSS as rapidly as possible; if your speed is less than 3 knots over the ground, motor.

In good visibility the islands will begin to appear as a seemingly arbitrary jumble of jagged humps along the horizon, very confusing at first and bearing little resemblance to what you are probably expecting. Initially you are only seeing the highest ground – closer-to they take on a more identifiable shape but remain confusing as the overlapping effect makes them appear as one continuous land mass rather than separate islands. However, once you begin to identify the conspicuous landmarks things rapidly slip into place.

Entry through St Mary's Sound

St Mary's Sounds lies between St Mary's and Gugh/St Agnes. Although nearly a mile wide, the navigable area is restricted by two groups of rocky shoals along its southern flank, the Spanish Ledges rocky shoals (least depth 0.9m LAT) and the Bartholomew Ledges (least depth 0.6m LAT), which effectively reduces the width of the sound to under two cables in places. Approaching from seaward the entrance to the Sound is marked to port by the unlit 'Spanish Ledges' BYB east cardinal buoy, and to starboard by the small iron lighthouse structure on **Peninnis Head**, which is fringed by low cliffs with distinctive and sculptural rock formations.

There is plenty of depth in the approach and in the sound but care should be taken if arriving near HW to avoid the **Gilstone** (dries 4.0m LAT) and lies four cables due east of Peninnis light. When the rock is submerged, if there is any sea running both it and the Gilstone Ledges further inshore are normally visible from the seas surging and breaking around them. Keep all of **Menawethan** (the easternmost of the Eastern Isles) well open of

Approaching Scilly from the east. At about eight miles off these low lying islands are only just beginning to show with St Martin's on the right, St Mary's, centre, and St Agnes level with the yachts shrouds

The red and white horizontal striped daymark is on the NE corner of **St Martin's**, and should not be confused with the all-white lighthouse on **Round Island**, at the NW corner of the islands. The pyramid shaped island of **Hanjague**, just east of St Martin's is also very distinctive, particularly in early morning sunlight. This and the rest of the **Eastern Isles**, soon begin to detach themselves.

St Mary's is most easily identified by the tall TV and radio masts (FR vert), the greater amount of trees and also a conspicuous white painted vertical wind generator on its eastern side. Approaching from the east, the sight of a prominent white lighthouse seemingly at the southern end of St Mary's can be confusing. This is not Peninnis light (a much more diminutive structure) but the old lighthouse on **St Agnes** immediately to the west of St Marys; from this approach the two islands look as if they are one.

There are two main entry channels into St Mary's Road when approaching from the east – the northernmost via **Crow Sound** requires sufficient rise of tide as you will have to pass over Crow Bar (least depth 1m LAT). This and the other shallows in the area tend to deter most first timers who usually opt for opt for the deeper, and better marked southern entrance through **St Mary's Sound**.

Peninnis Head and entrance to St Mary's Sound from SE

Newfoundland Point, the easternmost extreme of St Mary's you will pass to seaward of the Gilstone. Turn into the Sound once Peninnis Light is bearing just north of west.

Immediately west of Peninnis Inner Head the entrance to **Porth Cressa** is easy to identify – a long, wide inlet, with a sandy beach and houses at its head, and usually, if the weather is favourable, many boats anchored within.

From here the long straight line of the Garrison fortifications mirrors the line of low cliffs forming the coast. Close inshore, the 'Woolpack' YB south cardinal beacon

Isles of Scilly

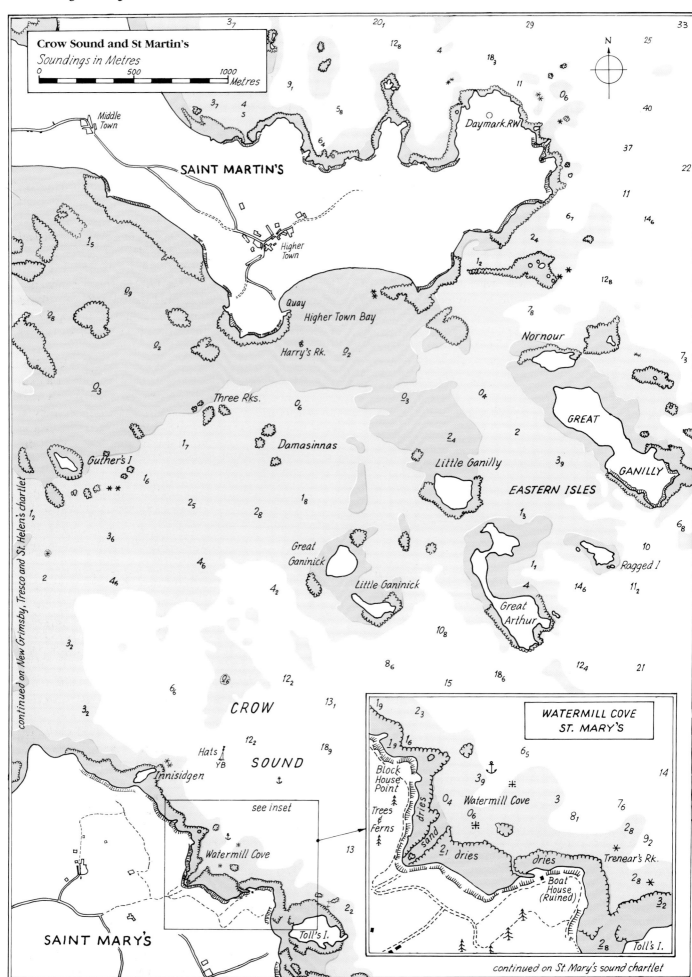

Crow Sound and St Martin's
Soundings in Metres

0   500   1000
Metres

Middle Town

SAINT MARTIN'S

Higher Town

Quay

Higher Town Bay

Harry's Rk.

Daymark.RW

Nornour

GREAT

GANILLY

EASTERN ISLES

Three Rks.

Damasinnas

Guther's I.

Little Ganilly

Great
Ganinick

Little Ganinick

Ragged I.

Great
Arthur

CROW

SOUND

Hats
YB

Innisidgen

see inset

Watermill Cove

SAINT MARY'S

Toll's I.

continued on New Grimsby, Tresco and St. Helen's chartlet

**WATERMILL COVE
ST. MARY'S**

Block
House
Point

Trees
&
Ferns

dries

sand

Watermill Cove

dries

dries

Boat
House
(Ruined)

Trenear's Rk.

Toll's I.

continued on St. Mary's sound chartlet

marking the Woolpack rock (dries 0.6m LAT) should be left on your starboard hand.

'Bartholomew Ledge' red can buoy (Fl R 5s) marks the inner end of St Mary's Sound on your port hand. If there is much ground sea from the west, as there invariably is, the sight of the swell breaking on the Bartholomew ledges can be intimidating, and with a wind of any strength against the tide, particularly from the east or south-east, a surprisingly steep, short and nasty sea can also build up in St Mary's Sound.

The ingoing (NW) tidal stream begins at HW Dover - 0310 the outgoing (SE) stream begins at HW Dover +0245 and attains a Spring rate of nearly 2 knots, and about three-quarters of a knots at Neaps.

From the Bartholomew Ledge buoy as long as you keep at least a couple of cables off the south and west side of St Mary's there are no hazards with the exception of **Woodcock Ledge**. Although this is covered 2.7m LAT, in common with most of the other submerged rocky ledges in the islands it can be a serious problem if there is any ground sea running, when the depth can be much reduced in the troughs, and dangerous if the seas are breaking. To avoid the Woodcock Ledge keep further out into St Mary's Road and do not begin to turn into St Mary's Harbour until the anchorage is well open and you can see the inner end of the quay.

Entry through Crow Sound

Crow Sound lies between the north eastern shore of St Mary's and south west of the Eastern Isles. Although wide at its mouth, any approach is conditional on having sufficient rise of tide to pass over Crow Bar which dries between 0.7m LAT and 0.5m LAT over much of its length but has a narrow channel, least depth 0.8m LAT, between its southern end and Bar Point on St Mary's. **Watermill Cove**, on the northern side of St Mary's, is a handy anchorage to wait for the tide, (see P 139).

Approaching from the east, if there is any amount of sea running do not cut the corner past the Eastern Isles but hold well to the south to avoid the shallower waters of the Ridge and Trinity Rocks. These can often kick up uncomfortable seas, and they break heavily in bad weather.

With the north-east going tidal stream which begins HW

Crow Sound from east: 'Hats' buoy, Innisidgen, left, and wooded Tresco in the distance

Dover +0500 a race develops across the entrance to Crow Sound for a couple of hours and this can extend up to two miles offshore. In reasonable weather this usually presents no problem but with a strong wind against tide it can be dangerous to small craft. Races also form to the east of the Eastern Isles, off Menawethan and Hanjague.

Within Crow Sound the streams are much weaker, rarely exceeding 1 knot at Springs, running (NW) into the Sound for 3 hours from HW Dover -0100, then out of the Sound (ESE) for eight hours, beginning HW Dover +0200 .

Once the Eastern Isles are well abeam on your starboard hand, you should be able to spot the unlit 'Hats' YBY south cardinal buoy about a cable ENE of Innisidgen, a low and jagged peninsula which rises to a distinctive conical and rocky point. Pass just south of the Hats buoy, but do not edge too close to Innisidgen as there are a couple of off-lying rocks (both drying 1.3m LAT). Instead, maintain a course parallel to the shore until the Crow Rock isolated danger beacon is well open of Bar Point, a long sandy beach backed by dunes, ferns and higher ground covered in trees.

You are now looking south-west down the full length of St Mary's Roads. **Tresco**, wooded and fringed with a long white sandy beach lies to starboard and in the far distance the two rounded green hills of **Samson** should be easy to spot. South Hill is the left hand of the two, and once Crow Rock beacon is centred on this you have the line of the deepest water over the bar.

Crow Rock coming onto transit with South Hill, Samson

Crow Rock is steep-to and can be passed on either side; from here on the depths begin to increase noticeably as you enter St Mary's Road. If beating, beware of the Pots (dries 1.8m LAT) and Round Rock (dries 1.5m LAT) which lie nearly half-a-mile off the southern shore of Tresco. When covered they can be avoided by either keeping Crow Rock beacon on a back bearing of 029°T, or ahead, by keeping the old lighthouse on St Agnes in transit with **Steval**, the low island on the distant westernmost point of St Mary's, to give a course of 209°T.

PASSAGE FROM MOUNTS BAY

A departure from Mounts Bay has the advantage of not having to worry about the tidal gate of the Lizard. However, to maximise on the tide it is probably best to push the foul tide out of Mounts Bay for an hour or so by leaving about 3 hours before HW Dover. By the time you reach the Runnel Stone it should be setting well to the SW, and later W for the next four hours, giving you a good shove on your way until 2 hours after HW Dover when it will be running northwards.

To be able to depart at this time you will have to be lying in Newlyn or anchored off Mousehole or St Michael's Mount as the tidal dock at Penzance does not open until 2 hours before local HW (HW Dover +0550). If you leave then you will have a foul tide for the next three hours and a strong tide setting you south for a further three.

PASSAGE FROM THE NORTH CORNISH COAST
Waypoints

Longships Lighthouse: 50°03′97N 05°44′85 W
Seven Stones Light float: 50°03′97N 06°04′28W
Round Island Lighthouse: 48°58′70N 06°19′33W

From the north Cornish coast, **Padstow** or **St Ives** are probable points of departure, with, weather permitting a landfall on the north-western side of the islands, and, most probably into **New Grimsby Sound**, (see Tresco section for approach and entry directions, P 147). It is approximately 70 miles from Padstow to New Grimsby, and slightly more to St Mary's. From St Ives the distance is about 40 miles.

If you are lying in the tidal basin at Padstow you can conveniently leave about an hour after local HW (HW Dover -0550) and you should then carry a fair tide along the length of the north Cornish coast and beyond for the next seven hours, after which it will become north going from 2 hours after HW Dover.

From St Ives Bay, again, leave an hour or so after local HW (HW Dover -0605) and you will have at least six hours mostly favourable tide. From Cape Cornwall the direction of the prevailing wind will most likely dictate your course to the islands, and whether you will pass to the north or south of the **Seven Stones**, a large group of dangerous rocky ledges (drying up to 2.9m LAT) nearly a square mile in extent 14 miles west of the Longships and seven miles NE of Round Island.

They are marked by a 12m high unmanned red light float, anchored two miles to the north-east. This can

provide a good check on the tidal set, but it is a lot less easy to spot than the lightship that formerly marked this reef which attained international notoriety with the *Torrey Canyon* disaster in 1967. Whether bound north or south of the Seven Stones, the whole area should be given a suitably wide berth.

As with the passage from the south coast, you will have to traverse the Land's End TSS, and my preferred route is to hold closer to the coast and make a departure from the mainland a couple of miles north of the Longships, with a course to pass well south of the Seven Stones, and a landfall on the NE side of the islands, where you can make the final decision on whether to enter by way of Crow Sound or St Mary's Sound or northabouts by New Grimsby Sound in the closer approach.

Entry through North Channel, Broad Sound and Smith Sound

As most visitors will be arriving from the mainland these westernmost approaches are the least likely to be used, and due to the number of off-lying dangers and strong tidal streams they are best avoided by newcomers to the area. If they are to be attempted, good visibility and weather are essential. Heavy overfalls occur in bad weather over the whole area north and south of the Bishop Rock lighthouse; numerous unmarked rocks and ledges, some just awash at LAT, create large areas of breaking seas in heavy weather or ground swell.

The approach to **Broad Sound** is made just over a quarter of a mile to the north of Bishop Rock lighthouse on a course of 059°T. The transit of the summit of Great Ganilly island just open to north of Bants Carn on St Mary's is extremely difficult to make out at this distance. Care must also be taken to allow for the tidal streams setting across your course. 'Round Rock' BY north cardinal buoy will be left well over on your starboard hand to clear the northern end of the Western rocks with the 'Gunner' YB south cardinal buoy to port. Both provide a good check on the tide which sets strongly across your track. You must then steer just to the north of the 'Old Wreck' BY north cardinal buoy, from where it is a clear run into St Mary's Road.

The **North Channel** brings you in to the west of Bryher and Samson and the Northern Rocks, which have the smaller islands of Maiden Bower and Mincarlo at their extremity. Ledges extend south-west from Mincarlo but the biggest hazard is the isolated Steeple Rock which almost dries LAT just over ½ mile south west of Mincarlo.

The normal approach from the north-west is made on 130°T, and here there are better marks, St Agnes old light house centred on the dip between the twin summits of the small island of Great Smith, although this, too, is not easy to see from a distance.

Smith Sound, though deep, is narrow and fringed by many unmarked rocky shoals making it a poor substitute for St Mary's Sound if approaching from the south. It should only be used with local knowledge.

Approach from the north: Round Island light, St Helen's and Men-a-vaur, right, with White Island, distant left

ST MARY'S

The largest and most fertile of the Isles of Scilly, St Mary's is also the most populated, its inhabitants variously dependent on farming, fishing, ship husbandry, pilotage and most recently, tourism. A blend of sheltered wooded and marshy valleys, tight clusters of daffodil fields surrounded by high evergreen windbreak hedges, and invigorating areas of bleaker, more windswept heathland, the island is criss-crossed with many deep hedged roads. Although it would seem impossible to get lost in such a small area, the virtual lack of signposts makes inland the forays very interesting!

The coastal footpath is one of the best ways to explore, the views are splendid and for ten gentle miles it winds past secluded coves, fine sandy beaches – frequently empty even in the height of summer – some memorable cliff and rock formations and a number of outstanding ancient burial chambers.

Old Town was the capital until Hugh Town began to develop after Star Castle and the garrison were built on the peninsula known as the Hue between 1593 and 1594, as part of Queen Elizabeth's continuing development of the chain of coastal fortifications begun by her father Henry VIII in anticipation of a Spanish invasion.

Scilly became an important Royalist outpost during the Civil War. Prince Charles, later Charles II was billetted in Star Castle for nearly six weeks during March and April 1646 while his escape route to the Low Countries was planned via the Channel Isles. In 1667, after 118 years of military occupation the army abandoned Scilly and the bemused inhabitants lapsed rapidly back into their hand-

St Mary's Harbour looking south nearing Low Water. Taylors and Newford Islands in foreground

133

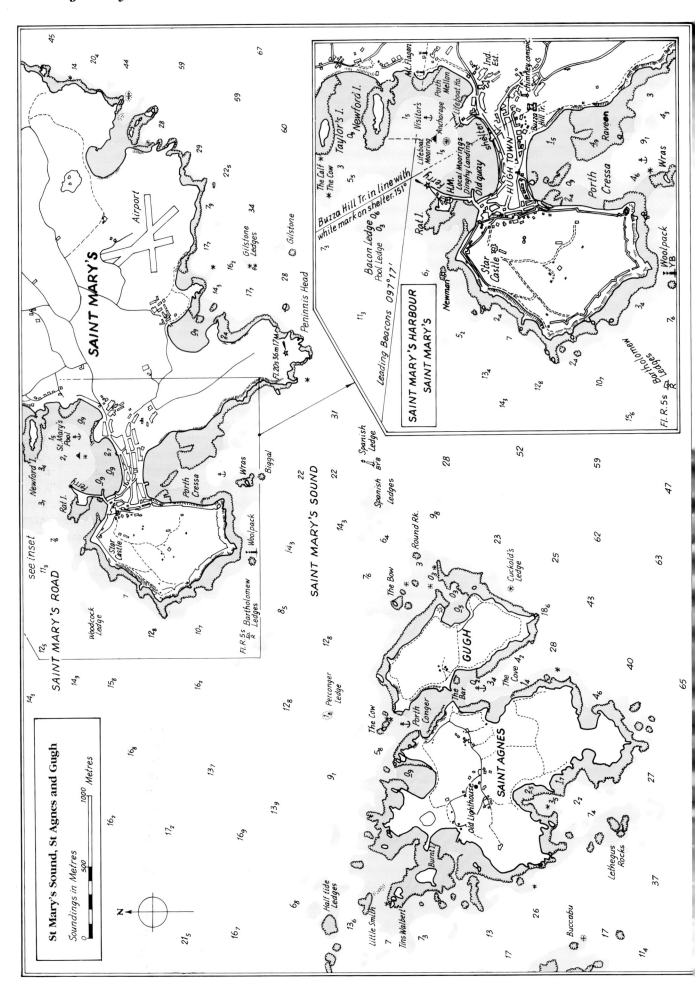

St Mary's Sound, St Agnes and Gugh

Soundings in Metres

SAINT MARY'S HARBOUR
SAINT MARY'S

SAINT MARY'S ROAD

SAINT MARY'S SOUND

SAINT MARY'S

GUGH

SAINT AGNES

to-mouth existence of fishing, farming, smuggling and reaping the God-sent rewards as the increasing amount of shipping resulted in ever greater numbers of shipwrecks – between 1745 and 1796, 750 men had died in the seas off Scilly from this cause alone.

The bay to the west of Hugh Town offered scant shelter and the Old Quay was built in 1601 to facilitate landing and give some extra protection to coasting vessels which dried out on the beach to discharge their cargoes. When Cowper visited the island in 1893 he commented on the need for a better breakwater linking the shore to the outlying Rat Island, in those days home to *an infectious diseases hospital containing eight beds'*.

The New Quay was built just after the turn of the century, not only linking Rat Island to the shore, but extending to seaward beyond, making it accessible at any state of the tide.

St Mary's has long been the base for the mainland link with the off-islands. In 1920 the islanders formed their own 'Isles of Scilly Steamship Company' and in 1926 the first *Scillonian* was built. Today the *RMS Scillonian III* carries on the tradition carrying visitors, vital supplies and the mail.

Additional cargo to the islands is shipped into St Mary's by the Steamship Company's other vessel, the pale blue *Gry Maritha*, from where it is distributed by smaller craft to the off-islands. Visitors can arrive either by sea or air but once in the islands the colourful fleet of large open motor boats belonging to the St Mary's Boatmen's Association becomes the primary form of transport. The Association was formed in 1958 and, remarkably, most of these sturdy waterborne charabancs were built prior to, during, or just after WWII; all are skipper owned and all the skippers were born and bred on St Mary's. The lack of a wheelhouse is no macho indulgence – their special brand of eyeball navigation precludes any such interference with vision. . .

Visitor's to St Mary's in their own boats have a choice of three commonly used **anchorages** depending on the prevailing weather, two of which are adjacent to Hugh Town, the main commercial centre of the Isles of Scilly. St Mary's Pool on the western side of Hugh Town is well sheltered from north-east through east to south, but becomes increasingly uncomfortable once the wind hooks much further to the west. **Porth Cressa**, on the south side of Hugh Town is sheltered in winds from west through north to east. **Watermill Cove** on the northern side of the island, though distant from any facilities, provides excellent shelter from south through to west, and is a particularly good retreat in strong south-westerlies.

As St Mary's possesses the only artifical harbour of any consequence in the islands it is not surprising that most first time visitors head for the anchorage in **St Mary's Pool**. However, there are several factors that combine to make this option less attractive than it might initially appear.

Hugh Town – St Mary's Harbour – is above all a commercial port, privately administered by the Duchy of Cornwall which has owned the Isles of Scilly since 1337. The facilities are often stretched to the limit and even more so with the ever increasing annual influx of visiting craft.

The demands on the near sacred ground of the Quay are particularly heavy – the outer end is used daily by the *RMS Scillonian III* between 1145 and 1630, and the various steps along its length are very busy prior to the 1015 and 1400 trip boat departures, and during the late afternoon when the boats return to disembark their passengers.

The quay is also much used by local fishing boats and off-island supply boats; neither their skippers, the boatmen or the Harbour staff take kindly to finding it or the steps impeded by yachts, the only exception being the innermost set of steps where the New Quay joins the Old Quay: these are reserved exclusively for visitors' dinghies.

If this is understood from the start a lot of potential problems can easily be avoided. If the Harbour Master, Terry Tyler, and tide permit, it is possible to go alongside for short stays of up to an hour or so on the inner end of the quay to shop or pick up crew. By prior arrangement with the Harbour Master, if there is space and no prevailing surge it is also possible to dry out overnight in the inner berths, particularly if you have a problem such as an engine breakdown. The bottom is firm sand but some agility or ingenuity will be required to get ashore at LW as there are no ladders.

Normally, when the outer end of the pier is free of commercial activity, boats are allowed alongside between 0830 (0945 Saturdays) and 1130 or from 1630 to 1700 to take on water and fuel (diesel only – petrol is available nearby in cans). Here again, there is but one simple rule of thumb to keep everyone happy – ask before you make any attempt to berth alongside, either contact the Harbour Office on VHF channel 16, call sign *St Mary's Harbour*, working channel 14, or hail the Harbour staff who are usually evident somewhere on the quay, and will always do their best to help.

The second problem is the harbour itself which does not always guarantee the shelter or comfort that most visitors are probably anticipating. The large inner part is shallow and dries as far as middle steps on the Quay at Springs. The remaining area of water sheltered by the Quay is entirely taken up with local moorings. None of these are to be used by visitors, even temporarily, and you should keep well clear of them – many of the moorings have very long floating pick-up warps just waiting to ensnare the unwary.

All visiting craft must therefore anchor north of the moorings, in St Mary's Pool. This is exposed to the west and extremely dangerous in strong winds from the north-west. It also suffers badly from swell if there is any ground sea running from the west or south-west (as there frequently is). The holding is indifferent, mostly sand and weed, with the odd rocky patch – to quote the Harbour Authority, '. . . *St Mary's Harbour is renowned for yachts dragging their anchors during gales, particularly when the wind is from the north-west . . . if the weather deteriorates when you are in the harbour think of seeking alternative shelter early'*.

St Mary's Pool is also the only place in Scilly where you will be charged for anchoring, (£6.50 up to 30 ft, £7.50 over 30ft) – the Harbour staff will usually be out to collect dues in the morning. However, in spite of all this, when the weather permits, the anchorage is usually very crowded which often constitutes a further problem in itself. Nevertheless, Hugh Town is the only shopping centre of any size within the islands and the only place where fuel and water are available alongside. A visit will probably be necessary during most stays in Scilly but personally I would only recommend it for a temporary daytime stop, or overnight if the weather is settled and calm.

St Mary's Pool – Approach from south

The direct approach from St Mary's Roads is partially encumbered to the west by the rocky and unbuoyed shallows of the adjoining **Pool and Bacon ledges** (least depth 0.3m LAT). Here again, if there is any ground sea running, the potential danger of this hazard is much enhanced by the surge and seas breaking upon it. If the seas are breaking white on Newford Island and the shore to the north of St Mary's Pool it will be very uncomfortable, and you should seek an alternative anchorage. Even if you are only intending to go alongside for water or fuel there will be a considerable surge along the quay.

There are two transits to clear Pool and Bacon ledges. If

St Mary's Pool: south passage leading marks in centre

approaching from the south-west you will use **the south passage** and you can begin to turn in toward the harbour once you have the two sandy beaches of Porth Thomas and Porth Mellon open of the outer end of the quay wall.

The beaches are separated by a scrub covered headland with a rocky foreshore and careful scrutiny (binoculars are a definite help) will reveal the two leading marks, the lower, a white triangle on a white post, and the upper, on the higher ground of Mount Flagon, a large X on a pole, normally painted white (but somewhat faded when the photographs were taken in July 1994), just to the right of a prominent bungalow. These give a leading line of 097°T. At night, each mark displays a FR light.

These are not the easiest marks to pick out from a distance, particularly in poor light or if the morning sun is behind them. The lower one is particularly elusive when there is a forest of masts in the anchorage which lies immediately to seaward.

St Mary's Pool anchorage

Give the outer end of the New Quay a wide berth to avoid the fairly constant stream of trip boats and everything else that seems inevitably to emerge from behind the harbour wall. If the *Scillonian III* or any other large vessel appears to be manoeuvring keep well clear. *There is a speed limit of 3 knots throughout the harbour.*

The visitor's anchorage is bounded by an imaginary line from the northern side of the lifeboat slip to the lifeboat mooring and across to the western end of Newford Island (see chartlet). Depths on the seaward side are about 2m LAT and reduce fairly steadily as you move inshore – select a suitable berth according to size and draught and the space available but *under no circumstances* should you lie to seaward of the lifeboat mooring or the large yellow Customs buoy. This area is used by the *Scillonian III* to turn into her berth – if you get in her way you will incur both the wrath of her Master and the Harbour Master. Take

When the weather permits, the anchorage in St Marys Pool is invariably crowded at the height of the season

Visitors are only allowed to berth alongside St Mary's busy quay to take on fuel and water at specified times with the Harbour Master's permission. Diesel is dispensed from the ingenious mobile pump in the foreground

care also to avoid the buoyed clusters of keep pots laid by local fishermen closer inshore along the fringe of Newford Island.

In south and easterly winds, if the anchorage in the Pool is overly crowded, Porth Loo, immediately to the north, is a viable alternative and also free of charge. Sound into a suitable spot midway between Newford Island and Taylor's Island, taking care to avoid the rocky ledges extending from both islands. The bottom is a mixture of sand, stone and weed and it is advisable to use a trip line.

St Mary's Pool – approach from north

Approaching from the north, particular care should be taken to avoid the **Cow**, an isolated rock (dries 0.6m) 1.5

Middle passage marks: Buzza Tower and white stripe on beach shelter roof

cables west of west of Taylor's Island which breaks if there is any amount of ground sea. The leading line of 151° T between the Cow and the Bacon Ledge, known locally as **the middle passage**, has the Buzza Tower as the upper mark, which is prominent and easy to spot on the skyline whereas the lower mark, a white painted beach shelter with a white stripe on the roof by the Town Beach, is much harder to see.

There is a third approach into St Mary's Pool known as **the north passage** inshore of the Cow and to seaward of the Calf Rock (dries 1.8m) off Taylor's Island, using a white painted mark of the eastern end of the Old (inner) Quay and a white painted window on the triangular shaped roof behind it. Unless you have seen these marks at close quarters and know exactly where and what you are looking for I would not recommend the north passage for strangers – given the small distance saved it is far easier to use the middle passage.

Facilities

Hugh Town is the only town in Scilly and though small it is a busy, bustling place during the season and even more so when the *Scillonian III* makes her daily appearance and the quay becomes a mass of frantic activity!

You can either land and leave your dinghy (on a good long painter) at the clearly marked steps by the inner end of the Quay, or alternatively anywhere on the beaches. This is often a far better ploy as the steps can become very crowded, and dinghies can suffer if there is any surge.

As already explained, water and fuel can only be obtained alongside between 0830 (0945 Saturdays) and 1130 or 1630 to 1700 by arrangement with the Harbour Staff. There is no longer a public tap on the quay, outside of these times containers can be filled if you ask at the Harbour Office which is situated upstairs and clearly marked at the seaward end of the Harbourside Hotel, the main building on the Quay. However you obtain it, you will be charged 5p a gallon for water.

Diesel is piped aboard by 'Sibleys' hand powered mobile pump which means they can deliver anywhere along the quay, although, power boat owners be warned, large quantities will take a long time. Normally they are on the quay during the morning period when yachts are permitted alongside the outer end, otherwise you can find them in the garage behind the Harbourside Hotel where petrol, oil, battery charging etc is also available, 0800-1200, 1300-1700 daily and 0800-1200 Saturdays.

Close by, in the area still known as Rat Island you will find Keith Buchanan, sailmaker, and Chris Jenkins' outboard repairs, beside which, stairs lead to the Harbour Authority's showers – two self-contained but rudimentary cubicles with shower, (£1 slotmeter) WC and washbasin, open daily between 0800 and 1645, and until 2100 during July and August – invariably you will have to queue. Rubbish skips are located behind the Harbourside Hotel and there is also a payphone and public toilets on the quay.

It is but a short walk along the quay and past the Mermaid Inn to find all normal provisions within easy reach of Hugh Street, the main centre. Early closing is Wednesday but most shops remain open during the season. They also tend to open earlier than on the mainland, usually between 0800 and 0830. The Post Office opens at 0815-1700, but closes at 1215 on Saturdays, and the Banks, of which there are only branches of Barclays and Lloyds are open weekdays only 0900-1600. Lloyds has the only cashpoint facility and be warned – both banks and the PO make a swingeing £5 service charge to cash cheques from other banks. There is a baker, two butchers, a 'Co-op' supermarket, open daily 0830-1730 and Sun 0930-1530, newsagent, chemists and several gift shops and a good mix of cafes, pubs and restaurants!

Somewhat hidden away behind the Isles of Scilly Steamship Office you will find their surprisingly well-stocked chandlers open 0800-1700 and 0800-1200 Saturdays; the 'Island Supply Stores', a hardware shop in Garrison Lane, is the only place in the islands where Calor and Gaz refills can be obtained.

The Tourist Information Centre is located at Porth Cressa beach; a synoptic weather chart and forecast are displayed here daily. Continue along the footpath past the Information Centre and public WC and you will come to that other cruising essential, the launderette, which does service washes only, Mon-Sat 0900-1300, 1330-1700. Next door there is a petrol station, and also a bicycle hire outlet.

If you follow the Strand you will find another chandler, 'Scillonian Marine', continue onwards and over the hill behind the lifeboat station and you will eventually reach the Porth Mellon industrial estate, which might seem to be an unappealing sort of place until you enter the portals of the 'Isles of Scilly Wholesale Company' which undoubtedly has the best selection of fresh produce on the islands and much else to offer in the way of provisions. If this is your only destination it is far quicker to land on Porth Mellon beach from the anchorage. Parr and Brown Marine Engineering is also based on the estate. There is a fairly limited choice of things to do ashore apart from the shopping and walks. In Church Street there is an excellent museum with the fully rigged gig *Klondyke* as its centre-piece; those less inclined to use their legs and partial to the eccentric can opt for one of the entertaining guided bus tours of the island, but undoubtedly the best spectacle is the Friday evening gig racing.

These elegant and racy clinker rowing boats are on average between 28 and 32 feet long with a beam of about five feet and were once unique to Scilly. They evolved during the last century to take pilots off to ships, either from the land or towed behind the larger sailing pilot cutters for use as boarding boats in the open sea.

Their fine turn of speed meant that they were inevitably used for less lawful activities. In an attempt to reduce the amount of smuggling the number of oars was limited to six by the Customs, whose boats they would otherwise outrun. Nevertheless Scilly gigs regularly rowed and sailed (with a fair wind they could set a lugsail and small leg o' mutton mizzen) across to Brittany to collect contraband. Their handiness and the skill of their crews was also crucial to the other major activity in Scilly, saving life and salvage from shipwrecks.

The revival of interest in these fascinating craft began in the 1950s when the Newquay Rowing Club bought most of the surviving gigs in Scilly, refurbished them and began to race them, including the oldest in existence, *Newquay*, built in 1812. It generated new enthusiasm for racing in Scilly, some of the boats were returned to the islands and new ones built, spurred on with annual visits by Newquay crews. During the last ten years, however, the whole sport has snowballed dramatically. New gigs and dedicated crews, both male and female, have emerged from just about every Cornish port, and there is a hotly contested calendar of racing events all around the coast with the highpoint, the World Pilot Gig Championship held in Scilly every Spring.

The weekly racing usually starts on the far side of St Mary's Roads off Nut Rock, and finishes off the end of the New Quay, which is an excellent vantage point. However, to get really close to the action join one of the many trip boats which follow the race. Packed to the gunwales with screaming and fanatical supporters, this is one Scilly experience that you will never forget!

Porth Cressa

When the weather permits – wind in the north-west, north or east and an absence of ground swell from the south – Porth Cressa can often be a far more pleasant alternative for a visit to St Mary's. Though sheltered in westerly winds, if there is much swell it can become uncomfortable, particularly around HW. Just as convenient for the town, and free of any commercial activity, this is inevitably a popular

Porth Cressa. The boats indicate the area of deeper water, Raveen is on their left, and the Wras on right.

anchorage and it can become overcrowded at times. If you are lying at the seaward end of the bay it is also a long dinghy ride to get ashore at HW, and often a scramble through rocks and weed at LW.

With any hint of a wind shift to the south-west, south or south-east or a sudden onset of swell from this direction, do not linger. Porth Cressa soon becomes untenable so seek an alternative anchorage at the earliest opportunty.

Approach and anchorage

Although the entrance to Porth Cressa opens immediately west of Peninnis Inner Head. If approaching from the east care must be taken to avoid Pollard Rock (dries 1.8m LAT) 100m due south of Inner Head and it is best to hold on more to the west before steering up towards the **Wras**, the isolated rocky island on the western side of the entrance.

If approaching from the west, keep well to the south of the Woolpack beacon to avoid a rock which dries 0.6m LAT just under a cable south-west of the Wras. **Biggal Rock** (2.4m) is a visible outlyer just south of the Wras, once this is abeam to port you can begin a gradual turn into the bay.

The appearance of Porth Cressa is *very* deceptive at HW – there are extensive drying rocks all along the eastern side, the head of the bay, and the area between the Wras and Morning Point dries completely on big tides. If possible, a low water approach has much to recommend it as all the hazards will then be evident, notably Fennel Rock (dries 1.8m LAT) which lies at the north-eastern corner of the large drying rocky base of the Wras.

Approach with care, keeping the Wras about 100m on your port hand and steering about NNW. Ahead, Raveen a 4.6m rocky islet marks the outer edge of Porth Cressa Brow (dries up to 4.9m LAT) on the east side of the bay; once this is abeam on your starboard hand you can begin to sound for an anchorage, the depth beyond this point rapidly reducing from 6m to 7m LAT to between 3m and 2m LAT. Brow Breeze rocks, which dry 0.3m LAT, extend across most of the head of the bay.

Although a number of underwater cables are indicated on the Admiralty chart these are well covered and not normally a problem. The bottom is mostly fine sand and the holding is again indifferent. I would not leave a boat unattended here on a single anchor for any length of time except in the very calmest of weather.

Watermill Cove

This pleasant and more remote anchorage at the northern end of St Mary's, just east of Innisidgen, can often be almost deserted if you are seeking somewhere less crowded. It is also handy if waiting for sufficient tide to cross Crow Bar, and invaluable as a bolt hole in strong south-westerly weather when St Mary's Pool becomes very uncomfortable and Porth Cressa is potentially dangerous.

In common with all the other anchorages in Scilly, Watermill has its limitations, and in winds of any strength much north of west it can become uncomfortable once the western reef covers unless you are of shallow enough draught to tuck right inshore. However, if there has been a blow from the south-west followed by the normal predictable wind veer to the north-west, it is not difficult to run back down under the lee of the eastern side of the island to the shelter of Porth Cressa, providing there is not too much ground sea from the south.

Watermill Cove is a less used but attractive anchorage which provides good shelter in south-westerly weather

Isles of Scilly

Although sheltered in southerly winds, swell will tend to work its way in from the east, particularly around HW, making it uncomfortable; once the wind shifts anywhere from south-east through east to north-west the cove becomes untenable and dangerous.

The coastal footpath to the west leads to the dunes and sandy beach at Bar Point, passing the Innisidgen burial chambers en route. The footpath to the east leads to the delightful beach at Pelistry Bay. Do not swim here, however, when the sandy spit to Tolls Island is covered as there are dangerous rip currents.

Watermill Cove has a fine sandy beach at LW, and from here it is a pleasant and easy 45 minute walk to Hugh Town. (Best route: take footpath from beach to lane, turn right at head of lane, continue past the duck ponds and follow road up hill to Telegraph, where there is a public pay phone, then first right once you have passed Telegraph tower. This leads down to Porth Loo - follow sandy footpath across head of Porth Thomas and Porth Mellon beach and join main road into town). Should you intend to return laden with provisions, Eric's Bus service will conveniently drop you back at the duck pond!

Approach and anchorage.

If approaching from the south or east, keep well to seaward to avoid the outlying rocks off Toll's Island and if approaching towards LW be wary of the outlying rock shown on the Admiralty chart as just awash LAT well to seaward in the centre of cove.

The cove is most easily located by the prominent pine trees on the fern covered hill overlooking the western side. Do not turn inshore until you have the sandy beach at its head on a south-westerly heading, but remember that at HW springs the beach is entirely covered. Alternatively, turn in once the distinctive sheer profile of Carn Wethers headland on the easternmost end of St Martin's is just open to the east of the conical island of Gt Ganinick to give a back bearing of 027°T.

Take particular care when approaching from the west. Keep up towards the 'Hats' YB south cardinal buoy and *do not cut the corner* off Block House Point, as the drying reef here extends a lot further to the north-east than you might imagine. Keep at least a cable offshore until the head of the cove is well open and you are onto the Carn Wethers back bearing. Nose your way in from here into a suitable depth – if you keep the highest, outer point of Innisidgen just open of the rocky islet off Block House Point you should have between 5m and 3m LAT.

Here the bottom is mostly sand and weed with some rock and the holding reasonably good; closer inshore there is a clear band of sand leading into the beach, fringed by weed covered rocks. Although the shelter improves closer to the shore, the rocks limit the swinging room.

Avoid the wide south-eastern part of the bay which is not only rocky but shallow, almost as far to seawards as the small boat moorings in the south-eastern corner. Here there is a ruined gig house and small slipway which has a cunning low tide, sandy approach channel dug through the rocks, making it the ideal place to land by dinghy if you are heading for Pelistry Bay.

PORT GUIDE: HUGH TOWN, ST MARY'S
AREA TELEPHONE CODE: 01720

Harbour Master: Mr Terry Tyler. Harbour Office, The Quay, Hugh Town, St.Mary's, Isles of Scilly. (Tel: 422768) Daily 0800-1700

VHF: Channel 16, working Channel 14, call sign *St.Mary's Harbour*, office hours

Mail Drop: c/o Harbour Office

Emergency Services: Lifeboat in St Mary's. Falmouth Coastguard

Anchorages: In specified area, St Mary's Pool, good shelter north-east through east to south, but subject to swell, often very crowded and holding poor. Dangerous in strong winds from west or north-west. Alternatives, Porth Cressa or Watermill Cove depending on weather

Mooring/berthing: No visitors moorings. Berthing alongside quay only by prior arrangement with Harbour Master

Dinghy landings: Where indicated at inner steps on quay. On beaches

Marina: None

Charges: Up to 30ft £6.50 per night inc VAT. Over 30ft £7.50 per night inc VAT

Phones: Public phone on quay behind hotel and several others in town

Doctor: Health Clinic (Tel: 422628)

Hospital: (Tel: 422392)

Churches: C of E, RC and Methodist in Hugh Town

Local Weather Forecast: Daily synopsis on board outside Harbour Office. Map and synopsis at Tourist Information centre

Fuel: Sibleys will pump diesel aboard alongside quay by prior arrangement with Harbour Master (0830-1130, Sat 0930-1200). Otherwise petrol and diesel in cans from Sibleys garages on the Quay and at Porth Cressa

Water: Pumped aboard alongside quay or in containers by prior arrangement with Harbour Master

Gas: Calor/Gaz from Island Supply Stores, Garrison Lane (Tel: 422388)

Tourist Information Office: Porth Cressa (Tel: 422536)

Banks/cashpoints: Lloyds (cashpoint) and Barclays in Hugh Street

Post Office: Hugh Street

Rubbish: Bins on quay behind hotel

Showers/toilets: Toilets on quay. Showers behind hotel

Launderette: Porth Cressa

Provisions: Most normal requirements available. Early closing Wednesday but most shops remain open in season

Chandlers: H.J Thomas, behind Isles of Scilly Steamship office, Hugh Street (Tel: 422710). Scillonian Marine, The Strand (Tel: 422124)

Repairs: H.J Thomas (see chandlers) T Chudleigh (Tel: 422505)

Marine engineers: Ask at Harbour Office or try H J Thomas (see chandlers). Parr and Brown (Tel: 422991). J Heslin (Tel: 422110). C.Jenkins (outboards) (Tel: 422539)

Electronic engineers: None

Sailmaker: Keith Buchanan, Rat Island Sail Loft on Quay behind Harbourside Hotel (Tel: 422037)

Transport: Daily ferries from St Mary's to Penzance (except Sundays) British Airways helicopter service to Penzance. (Main line British Rail connections at Penzance), Skybus air connections to Exeter Plymouth, Land's End and Newquay airports

Car hire: Sibleys (Tel: 422431)

Yacht Club: Isles of Scilly YC (Tel: 422352)

Eating out: Good selection from fish and chips to pubs/restaurants/bistros

Things to do: Museum. Fine walks/beaches. Gig racing every Friday evening

ST AGNES AND GUGH

'**W**ith this Mark you run in amongst many rocks terrible to behold . . .' still serves as a warning to those who have the good fortune to spot the old lighthouse on St Agnes from afar. Closer to, it is rarely long out of sight, standing proud on the low summit of this gentle island, as it justly deserves, for this was the earliest offshore light to be established by Trinity House.

The coal braziers were fired up for the first time on 30th October 1680 and they, and the Argand lights that succeeded them continued to burn bright until 1911 when the lighthouse was deemed redundant after the building of the light on Peninnis Head. Although the tower and its keepers house are now a private home leased from Trinity House, they are still responsible for keeping this important navigational daymark gleaming white.

Although St Agnes and Gugh outwardly display similar physical characteristics to the rest of the inhabited islands, there is one fundamental if underlying difference – they were never joined to the others which were originally a single land mass, until the sea levels rose during the Bronze Age. Leaving the waters of Scilly shallow and rock strewn, it also gave rise to the romantic notion that these are the last remaining fragments of the lost land of Lyonnesse which once, supposedly, extended far west of Land's End. Even today it is theoretically possible, on big Spring tides (in practice you would not have enough time), to make a low water circuit of most of the islands on foot, crossing via Crow Bar to Tresco and St Martin's, and across New Grimsby Sound to Bryher and Sampson.

The inhabitants of St Agnes and Gugh form the most

Porth Conger. The wind has just shifted to the north-west and is already sending in a bit of swell. The distant boats beyond the sand bar are anchored in the Cove and are now enjoying much better shelter

south-westerly community in the British Isles, and like the rest of the islands their livelihood today is dependent on flower farming, a small amount of fishing and the tourist industry. A century ago, the men of St Agnes were renowned worldwide as among the finest pilots in Scilly, capable of navigating ships from the western approaches as far north as Glasgow or as far east as Bremen.

This compact and attractive island is easy to explore in little more than a couple of hours. Its central part is a dense and intimate patchwork of flower fields surrounded by high windbreak hedges, in marked contrast to the open moorland of Wingletang Down to the south. Here, in an inlet on the rocky and boulder strewn coast, you you might find terracotta beads from a wrecked Venetian trader in the sands surrounding Beady Pool.

Castella Down lies to the west and here it is the Troy Town stone maze that will catch your eye. Laid out on the low cliff top by the bored son of the lighthouse keeper in 1729, it also provides an impressive if chilling panorama of a very different kind of maze, the off-lying rocks bordering Smith Sound, Annet, the western rocks and the distant finger of Bishop Rock lighthouse. This was completed between 1852 and 1858 by Nicholas and James Douglass to the design of James Walker with a team of workmen who were billetted on the small island of Rosevear, where the remains of their cottages can be seen.

Standing in one of the most exposed locations in the world, in 1874 the lighthouse was hit by such heavy seas that the lenses were broken, the structure was felt to *'reel and stagger'* and the upper gallery was filled with sand. By 1887 the foundations had undergone major strengthening and the height increased by forty feet to 167 feet (50.9m) making it the tallest lighthouse in Britain. In spite of this, in heavy gales, much of it is frequently hidden by the mountainous seas.

Periglis Bay on the north-west corner of the island provides a small natural but drying harbour for the local boats, overlooked by the church and old lifeboat house which closed in 1920.

Apart from two houses more reminiscent of somewhere on the eastern seaboard of Maine, the featureless island of **Gugh** is connected to St Agnes by a sand bar except at HW Springs, a heathy, heathery place with many rocky outcrops and several important megalithic remains, including the Old Man of Gugh, a nine foot standing stone, and Obadiah's Barrow. Today, some of the island's present inhabitants are likely to make a more immediate impression, as the southern end is home to a large colony of very aggressive gulls during the nesting season in early summer.

There are two possible **anchorages** depending on the direction of the prevailing wind and sea and the sand bar connecting Gugh to St Agnes is common to both. This dries about 4.6m LAT, and covers on Springs when there is a noticeable tidal stream from north to south through the inlet, making bathing dangerous.

Porth Conger on the northern side of the bar is sheltered from north-east through south to west, but exposed once the wind edges any further north. **The Cove** on the southern side is well-sheltered from west through north to north-east, except around HW if the bar is covered when

fresh northerlies and north-westerlies can create quite a bit of chop. Although marginally sheltered in south-westerlies and easterlies, any swell will make itself increasingly felt, particularly towards HW. In the onset of winds from the south or south east, get out as soon as possible.

Approaches and anchorage, Porth Conger

Approaching from the east keep to the north of Spanish Ledges until the Bow (10m), an isolated island off the eastern side of Gugh, is abeam when it is safe to steer across for Kittern Rock (17m) on the northern end of Gugh. Keep about 100m off the rock and steer for the northern end of the Cow, leaving it several boat lengths to port to avoid the ledge extending to the north-west, before bearing round into Porth Conger, and making an approach midway between Gugh and St Agnes. There is a short cut used by local boats in the narrow passage between the Cow, the Calf (dries 1.2m) and the shore which is best avoided by strangers.

Approaching from St Mary's, keep well to the west of the Bartholomew Ledges buoy and the North Bartholomew shoal which breaks heavily if there is any sea running. So too does the **Perconger Ledge** (least depth 1.8m LAT). If the visibility is good, the best plan is to use the leading line for St Mary's Roads – the daymark on St Martin's over the top of Creeb Island until you can see the gap opening between Gugh and St Agnes. In the distance the Hakestone (2m) should be visible beyond the sand bar – if you keep this more or less on the centre of the bar it will give you a clear line in.

The deepest water lies north of the end of the jetty. Sound into a suitable depth and anchor well clear of the few local moorings and the approach to the jetty which is in regular use. Close to the St Agnes shore there is a rocky patch, otherwise the bottom is clean fine sand, and the holding indifferent. South of the jetty, in what looks to be the ideal and most sheltered place to anchor, depths reduce rapidly, and much of this flat sandy area dries at Springs, making it ideal for bilge keelers.

Approach and anchorage, The Cove

The approach to The Cove will invariably be made from the east or north east. If arriving from St Mary's the safest bet is to leave St Mary's Sound to the north of Spanish Ledge buoy and then steer south until Pidney Brow, the 13m hill on the southern end of St Agnes, is just open of the Hoe, the prominent rock on the southern tip of Gugh. This line keeps you in clear deep water past **Cuckolds Ledge** (dries 1.4m LAT). As you close the Hoe, which is steep-to, keep a couple of boat lengths to seaward and once the Cove opens and you have the distant lump of the Cow lined up on the centre of the sand bar, you can steer straight up into the anchorage, leaving the Hakestone (2m) a 50m or so on your starboard hand. The only real hazard in the approach is the Little Hakestone (dries 3m LAT) on the south western corner of the entrance – as long as you keep up to the Hakestone this is not a problem, the entrance is over 300m wide and there is plenty of room.

It is possible, in clear, calm weather to take a short cut to the south of the Spanish Ledges once you have got your

The Cove viewed from Gugh at half-tide. Cove Vean is below the old lighhouse; note weed on the rocks in foreground

local bearings, by using the transit of Steval, the islet off the westernmost extreme of St Mary's in transit with Hangman Island in New Grimsby Sound. This gives a back bearing of 344°T but takes you close to Round Rock (dries 1.2m) and the Brow Ledge which extends east from Gugh. Do not be tempted to cut the corner but keep on this course until Pidney Brow is well open of the Hoe.

There will invariably be other boats at anchor in the Cove, sound in towards the head of the inlet and select a suitable spot. The depths reduce gently the further north you proceed with a mostly sandy/weed bottom and reasonable holding. Ideally let go somewhere along the centre line of the inlet as the shoreline is very rocky on both sides (but easy to spot from the weed), except across the head of the bay where the bar forms a fine sandy beach that is excellent for swimming. The bar should be avoided when covered as the current can run hard across it.

Avoid Cove Vean the inlet on the western side, which is rocky and dries almost to its mouth on big tides. Although the Admiralty chart shows a number of underwater cables running out of the Cove they are well buried and not a problem.

Facilities

If anchored in Porth Conger, do not leave dinghies tied alongside the jetty, which is busy with pleasure boats often late into the evening, but land on the sandy beach/old slip immediately below the 'Turk's Head Pub', a particularly convivial watering hole where bar food and evening meals are available.

If anchored in The Cove land by the St Agnes end of the Bar where the footpath leads up to join the concrete road. Turn right for the pub, turn left and you will climb the gentle hill past two cafe/tea rooms and eventually reach the surprisingly well-stocked Post Office, General Store and Off-Licence, which can cater for most normal provisioning requirements - be warned, there is usually not a surplus of bread. If you are lingering here for a while they will order this (and anything else you might need) from St Mary's on a daily basis.

Continue along the road toward the old lighthouse and you will find a public phone box; there is also a payphone in the Turk's Head. There is no water or fuel available.

TRESCO, BRYHER and SAMPSON

Those seeking expensive works of art or bottles of claret, exotic shrubs and flowers or merely an attractive and sheltered anchorage need look no further than Tresco. The second largest island in Scilly, it is private and unique in that it is leased from the Duchy of Cornwall by the Dorrien-Smith family, the descendants of Augustus Smith, a Herefordshire land owner who took over the administration of the islands in 1834. Assuming the title of 'Lord Proprietor of all Scilly', his beneficial dictatorship lasted a remarkable 38 years until his death in 1872 and transformed their prosperity.

Basing himself on Tresco, he built the house known as Tresco Abbey as his main residence and immediately began to establish the Abbey Gardens for which the island is internationally renowned. Today, the 17 acres of south facing terraces are a botanist's paradise, with sub-tropical plants from all over the world flourishing in this unique outdoor location. Crew members less interested in flora can ponder sagely on the collection of ship's figureheads and memorabilia salvaged from local shipwrecks housed in the *Valhalla* museum within the gardens.

On a wider scale Smith did much to enhance the general well-being of the inhabitants, and eventually managed to virtually wipe out the rampant smuggling and establish a well-ordered society. A fervent believer in education, he introduced the first compulsory schooling in the United Kingdom, established five shipbuilding yards, and created a new industry with the cultivation of early flowers and

New Grimsby Sound looking south-east from Bryher. Hangman Rock is in the foreground with anchored boats behind and the line of visitors buoys is clearly visible along the far Tresco shore

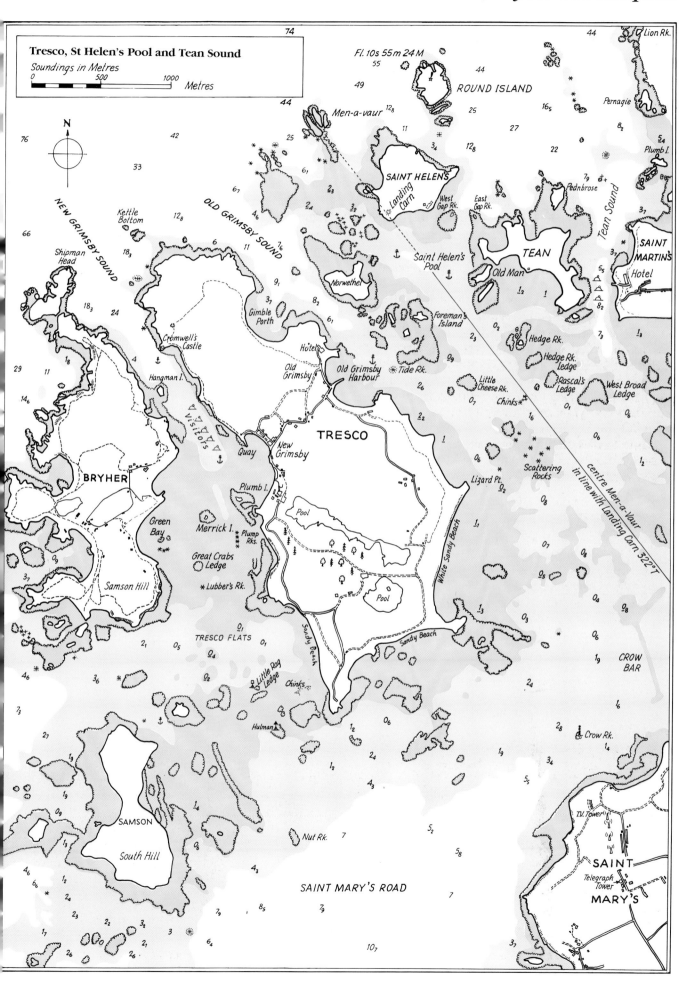

Tresco, St Helen's Pool and Tean Sound

Soundings in Metres

0   500   1000   Metres

74

Fl. 10s 55m 24M

55

49

44

44

ROUND ISLAND

25

16₅

44

Lion Rk.

Pernagie

8₂

Plumb I.

76

42

33

Men-a-vaur

12₈

11

3₄

12₈

27

22

SAINT HELENS

5₂

7₉

Pednbrose

8₃

66

Kettle Bottom

18₃

12₈

OLD GRIMSBY SOUND

6

11

7₆

4₆

2₃

2₄

3₂

West Gap Rk.

Landing Carn

East Gap Rk.

Saint Helen's Pool

TEAN

Old Man

1₂

1

5₅

3₇

SAINT MARTINS

Hotel

8₂

1₃

Shipman Head

18₃

24

Cromwell's Castle

9₁

8₂

6₁

Norwethel

3₇

Gimble Porth

Foreman's Island

2₃

2₉

Hedge Rk.

Hedge Rk. Ledge

Rascal's Ledge

West Broad Ledge

NEW GRIMSBY SOUND

29

11

1₈

Hangman I.

Old Grimsby

Hotel

Old Grimsby Harbour

Tide Rk.

2₆

Little Cheese Rk.

0₇

Chinks

0₁

0₆

14₆

4

Quay

New Grimsby

2₂

1

Scattering Rocks

1₆

0₆

1₂

BRYHER

Visitors

TRESCO

Pool

0₆

0₈

3₇

Green Bay

Plumb I.

Merrick I.

Plump Rks.

Pool

Lizard Pt.

0₂

1₁

0₇

0₈

Samson Hill

Great Crabs Ledge

Lubber's Rk.

Pool

White Sandy Beach

0₅

0₄

0₈

2₁

0₅

0₁

TRESCO FLATS

0₄

0₂

Sandy Beach

Sandy Beach

1₃

0₃

0₆

CROW BAR

1₉

4₆

3₆

Little Rag Ledge

Chinks

Hulman

1₂

0₆

2₄

1₆

7₃

2₇

Crow Rk.

2₈

1₄

1₂

2₄

4₃

5₅

3₄

1₉

1₃

SAMSON

1₄

T.V. Tower

5₂

Nut Rk.

7

5₈

0₉

South Hill

0₆

7

SAINT

4₆

6₆

1₂

4₃

SAINT MARY'S ROAD

Telegraph Tower

2₄

2₃

2₂

3₄

3

7₉

8₅

7₉

MARY'S

1₇

2₆

2₆

2₁

6₄

10₇

3₇

centre Men-a-vaur in line with Landing Carn 322°T

145

potatoes which have long been one of the commercial mainstays of the Isles of Scilly.

Tresco's economy today is based on the hyper-efficient management of tourism, and it even sports its own heliport with direct flights from Penzance. Although there is still a small amount of farming, the majority of the old workers cottages are now converted into up-market self-catering holiday homes and timeshare properties. The luxurious Island Hotel, overlooking Old Grimsby on the eastern side of the island, is even more prestigious and would not look out of place in a Caribbean setting.

Apart from a few tractors and golf buggies Tresco is blissfully car free, although the few miles of road seem at times to be positively overrun by small children on hired bicycles, of which there are nearly 300! It is, however, a quite intriguing island of two very distinct halves, both physically and spiritually.

The south-eastern end is low lying, heavily wooded in the vicinity of the Abbey Gardens, with two large reed fringed lakes, Great Pool and Abbey Pool, and a shoreline of magnificent white sandy beaches. The cottage lined road that climbs the gentle hill between New Grimsby on the west side, Dolphin Town in the centre and Old Grimsby on the east forms a natural divide. North of this, delightful walks along the coastal footpath soon take you back to nature in the raw, the untouched, windswept expanse of rock strewn heathland, gorse and heather of Castle Down, named after the gaunt granite remains of King Charles Castle. Built between 1550 and 1554, this was superceded in 1652 by Cromwell's Castle which was far better sited to protect the entrance to New Grimsby.

Walking the rugged cliff top around the north-eastern end of the island, you can search for, and hopefully find, the island's best hidden attraction, Piper's Hole, a cave reached by a steel ladder leading down into a gully. At the head of the gully the narrow rock encumbered entrance to the cave eventually opens into a much larger inner chamber where there is a shallow pool, its bottom composed of a horribly glutinous clay. A torch, some candles and a complete lack of claustrophobia are essential for the successful completion of this expedition!

For the visiting yachtsman, Tresco is best known for having the most protected anchorage in Scilly, **New Grimsby Sound,** which lies between the western side of the island and Bryher, a long narrow passage, for the most part little more than a couple of cables wide. Here, too, the Tresco Estate has 10 visitors' moorings available for hire; in high season the demand on these is always heavy and the anchorage inevitably crowded, as it is the ideal base from which to explore both Tresco and Bryher.

New Grimsby Sound is perfectly sheltered in winds from south-west to west, and north-east through to south-east. Although safe, in strong winds from the north, particularly north-westerlies, and strong southerlies it can become very uncomfortable, and rough with wind against tide, particularly at Springs. Nor too, is it entirely devoid of swell and like just about every other anchorage in the islands it can be quite rolly at times.

Old Grimsby Harbour on the eastern side of the island is the best alternative anchorage on Tresco, and although shallower and more exposed, in fine weather it is often quieter and less crowded. Depending on your position there is good shelter in most wind directions except east and it is prone to swell in northerlies.

New Grimsby Sound – approaches and anchorage

Given favourable weather the easiest approach to New Grimsby Sound is from seaward, by way of Round Island and the northern side of the islands. The soundings drop away fast and there are no offshore hazards more than a few cables from the main land masses; however, if there is much wind or ground sea running from the west or south-west, the sight of seas breaking heavily along the rocky north-western shores can be somewhat daunting. The tidal streams generally run at up to 2½ knots at Springs (and reputedly over 4 knots close inshore off Round Island) creating confused seas with wind against tide for several miles to seaward; close to the entrance to the sound overfalls can

New Grimsby Sound, northern approach: Cromwell's Castle, Hangman Rock and anchored boats within

New Grimsby Sound, southern approach nearing High Water. The Hulman beacon is on the right, Little Rag Ledge beacon is just visible left of centre and Hangman Rock is distinctive in the distance

occur off Kettle Bottom Ledge from 3½ to 4½ hours after HW Dover.

The southern approach to New Grimsby from St Mary's Road is across **Tresco Flats**, an extensive area of drying sands, which seems both shallow and complicated at first glance. As long as you do not attempt it on a falling tide and preferably after half flood, it normally presents no problem for boats of average draught. At MHWS Springs you should easily carry 5m over the Flats and about 3.3m at MHWN.

Approach from north.

Once Round Island is well over on your port quarter, and the rounded humpback of St Helen's Island and the distinctive summits of Men-a-vaur are abeam, if you hold a south-westerly course about a quarter of a mile from the shore, the northernmost point on Bryher, **Shipman Head**, will appear as a long, undulating and rocky promontory on your port bow. This forms the western side of the entrance to New Grimsby Sound, keep your distance off until you can see clearly into the Sound and identify the two prominent marks within: **Cromwell's Castle**, a flat topped rounded stone tower is on the eastern side of the entrance; the pyramid shaped bulk of **Hangman Island** (16m) lies slightly further into the sound on the western side.

Though relatively narrow, the entrance to the Sound is deep and easy to enter as long as due care is taken to avoid **Kettle Bottom Ledge** (dries 3.2m LAT) which extends two cables to the north-west of Tresco, forming a trap for the unwary who try to cut the corner. Normally, if there is any hint of ground sea it is easy enough to spot from the ominous surge over it.

Once the steep western side of Hangman Island is bearing about 157°T it is safe to steer inshore, but make due allowance for the tide which sets across the entrance in excess of 2 knots at Springs, the stream running east at HW Dover +0115, and west HW Dover -0510. Shipman Head is

steep-to, and it pays to keep up to this side of the entrance, before easing midway between Cromwell's Castle and Hangman Island in the final approach.

Immediately south of Cromwell's Castle there is a deep water anchorage with depths ranging between 7m and 11m LAT, mostly sandy bottom. Once Hangman Island is abeam the depths rapidly reduce to between 3m and 4m. On the Tresco shore there is a prominent rocky outcrop, beyond which the 10 visitors' moorings extend southwards along the eastern side of the sound. If there is no indication that they have been reserved and you are happy to pay the £7.50 overnight charge merely pick one up – they are all good for up to 15 tons, maximum 15.2m (50 ft) and no rafting is permitted. Should you wish to try and reserve one in advance you can contact the Harbour Master Henry Birch (Tel: 01720 422857. Normally he will be out to collect your dues in the late afternoon.

If you prefer to anchor, either let go to the north of the moorings, clear of them along the Bryher side of the channel or further to the south, but here you must take care, beyond the cluster of private moorings the depths reduce rapidly and there is little more than than 0.5m LAT abeam of New Grimsby Quay. *There are also two underwater power cables, clearly marked with triangular yellow boards on both the Tresco and Bryher shore, which must be avoided.*

Shallow draught boats can work their way closer inshore but will not be able to dry out as anchoring is not permitted anywhere inside the low water mark on Tresco, and definitely not inshore of a line from the end of the Quay south to Plumb Island.

The quay is in constant use with trip boats and ferries and their approach must not be impeded. Visitors are not allowed to lie alongside the quay except in special circumstances with the Harbour Master's permission. Dinghies should be left on the sandy beach and ***not*** alongside the quay steps.

Due to the restricted swinging room, if there are many boats at anchor, the convoluted tidal streams can often cause problems. At Springs they can run at up to 2 knots from 1 hour before local LW to 1 hour after LW (HW Dover -0140 to +0140)), and the flow is intriguing, as the direction changes four times every twelve hours running south-east HW Dover -0010, north-west HW Dover +0125, south-east HW Dover +0415, north-west HW Dover -0340.

Approach from south over Tresco Flats

For boats drawing no more than 1.2m this approach can be made with care from about 2 hours after LW; deeper draught boats should wait at least until half flood. Approaching from the direction of St Mary's leave through the south passage and steer 340°T across St Mary's Road. The island of **Samson** with its twin rounded hills lies well away on your port bow, with **Bryher** beyond and the larger wooded bulk of **Tresco**, with its long white sandy beaches to starboard. The isolated **Nut Rock** (1.5m) is a useful and fairly visible outlier to the Tresco channel – leave this a couple of boat lengths to port. From here the **Hulman Beacon** – a rather spindly looking iron perch with a green triangular topmark – will be visible on your starboard bow, steer a course to leave this a good 30m on your starboard hand as the rocks extend well beyond the perch.

On your port bow the **Little Rag Ledge Beacon** is another spindly affair, topped with a red painted radar reflector. Do not steer directly for this but make a gradual turn towards it to avoid the southernmost edge of the reef (dries 1.3m LAT). Take care not to stray too far to starboard either, to avoid the isolated patch of the **Chinks** rocks (dries 1.3m LAT).

It sounds and looks more complicated than it really is. The water is disturbingly clear and as long as the light is good you will not only be able to see the darker areas of weed over the rocks and the clean sand beneath your keel but even the crabs running around upon it!

Once Little Rag Ledge Beacon is abeam, with enough rise of tide you can proceed directly across the 'Flats', the large sandy spit which extends nearly ¼ mile westwards from Appletree Point on Tresco, and dries up to 1.7m LAT in places.

In the far distance you should now be able to spot two rocky islands in mid channel – the left hand one, **Merrick Island**, is the lowest (2.6m), **Plumb Island** (7m) is on the right. Further beyond, the pyramid shaped profile of **Hangman Island** (16m) is quite unmistakeable. Keep Merrick in front of Hangman, on about 340°T, and you will avoid the rocky ledges closer inshore off Appletree Point which marks the northern end of the long sandy beach backed by dunes on your starboard hand.

However, if you are doubtful whether you have sufficient depth over the Flats, deeper water will be found by swinging more to the west once you are past Little Rag Ledge beacon. Ahead, Samson Hill on the south end of Bryher has two small summits, steer for the left hand one on about 302°T for just over ¼ mile. Further away on your port hand Yellow Rock is a small rocky islet midway between Samson and Bryher – once you have this abeam

and bearing about 230°T you can turn north-east. In good visibility the top of the distant Bishop Rock lighthouse over the centre of Yellow Rock gives a good idea of when to turn.

The bottom is all sand so even if you do touch, with a rising tide you will come to no harm. The only real hazard is the isolated Lubbers Rock (dries 1.7m LAT) which should left on your port hand: as long as you can see the southern end of the more distant island of Mincarlo open of Works point, the southernmost tip of Bryher, you will clear this safely.

Beyond this point the Bryher side of the channel is fringed by Little Crab Ledge (dries 2.4m LAT) and Great Crab Ledge (dries 5.3m LAT) and only just covers at HW – normally some of it can be seen as cluster of rock heads. On the Tresco side the Plump Rocks dry 2.2m LAT. As long as you keep the small white gabled building on New Grimsby Quay just open of Plumb Island, you will pass up the centre of the channel clear of all these hazards. Steer between between Plumb Island and Merrick, where the deepest water will be found on this side of the channel, and once Merrick is abeam steer straight for the lower end of the moorings to avoid the only other hazard on the Bryher side of the channel, Queens Ledge (dries 2.5m LAT).

Old Grimsby Sound

Given suitable weather and an absence of ground sea, the easiest approach to this alternative Tresco anchorage is from seaward. Ideally, aim to make your approach soon after LW when most of the hazards will be easy to spot. Do not cut the corner off the isolated rounded rocky island known as Golden Ball if approaching from the north-east as there are several isolated off-lying rocks.

As with New Grimsby, hold a course ¼ mile to seaward until you can see clearly into the Sound – the island of **Norwethal** on the eastern side is a good mark as it has a very distinctive flat topped rock formation at its highest point – and make your approach following the line of the Tresco shore, making due allowance for the tide which can set strongly across the entrance.

Personally I would suggest visiting New Grimsby first, and reconnoitring the entrance to Old Grimsby from the land. It is a then but a short hop around the northern end of Tresco, taking care, of course, to avoid Kettle Bottom.

Once inside the entrance to the sound Little Kittern Rock (dries 1.9m LAT) is the biggest hazard, and lies on your port hand opposite the entrance to Gimble Porth, the sandy bay on your starboard hand, which can provide a pleasant temporary anchorage in the north-eastern corner. Merchant's Point is the rocky headland at the southern end of Gimble Porth, aim to pass 30m or so off this and maintain this offing. In the distance, on your starboard bow, the prominent ruined block house overlooking the southern end of **Green Porth** will soon be in sight on a grassy headland, with the long sandy beach and dunes of Green Porth running round to the small pier that separates it from Raven's Porth.

Once this is well open to starboard, you can anchor anywhere in mid-channel where there are depths of just over

3.0m LAT, with a sandy/weedy bottom, ideally somewhere abeam of Middle Ledge and clear of the local moorings. The tidal stream can be strong here, up to 2 knots at Springs, running south-east through the sound for eight hours beginning HW Dover -0210, and north-west for the remaining four hours.

Shallower draught boats can edge out of the tide and closer into Green Porth where there is generally better shelter and less swell if any is prevailing. Do not anchor inside the LW mark, and take care to avoid the underwater cable which is indicated by the sign at the head of the beach, and runs out of the bay to the south of the Quay.

Leave your dinghy either on the beach or the Quay; the long landing causeway at the northern end of Raven's Porth belongs to the hotel and is is private.

If you continue any further south, take care once Trafford Rock is abeam as the isolated trap of **Tide Rock** (dries 1.4m LAT) lurks 100m to the SSW. To avoid it, bear over towards Block House Point once Middle Ledge is abeam.

Passage from Old Grimsby to St Mary's Road

With sufficient rise of tide it is possible to exit from Old Grimsby to the south and into St Mary's Road by following the Tresco shoreline beyond Block House Point. Cooks Rock (dries 4.3m) which lies about 150m due east of the point is the only hazard; from here on keep about 250m from the dunes backing the white sandy beaches until you are past Rushy Point which has a post with a diamond shaped yellow sign marking an underwater cable. From here you should be able to see Crow Rock beacon and beyond, on St Mary's skyline, the Telegraph Tower. Keep these in transit to give a course of 162°T and you should avoid all the rocky ledges off the south-eastern end of Tresco; as most are likely to be covered, a good lookout on the bow is advisable. From Crow Beacon, bear away for St Mary's, keeping Steval and the old lighthouse on St Agnes in transit, to give a course of 208°T.

Facilities.

Tresco Estate, and in particular, their Harbour Master do everything they can to assist the visiting yachtsman. For the past 30 years and more this was the cheerful lot of Laurie Terry who retired in 1994; hopefully his successor, Henry Birch will continue in the ever-friendly and helpful manner that was always Laurie's hallmark.

Water containers can be filled free of charge from the tap by the Quay Cafe, where there are also public toilets. There is no diesel available but petrol can be obtained in cans if you ask at the estate office, which is in the large complex of farm buildings on the southern side of the bay. Rubbish must be left in bags in the trailer behind the quay shop.

The Abbey Gardens, which are open daily 1000-1600 with a moderate admission charge, are a mile and a half away. The 'New Inn', the only pub, sports a bar that was completely refurbished with prime pitchpine planks washed ashore during the winter of 1993. There is a good range of bar meals and basic showers (£1) if you enquire at the bar.

Old Grimsby Sound looking north from Tresco. Norwethal is distinctive near the entrance and St Helen's is in the far distance with Round Island light peeping over its summit. Note Tide Rock just showing in channel on right

Isles of Scilly

A short distance up the hill the Tresco Stores, Off-licence and Post Office has a comprehensive selection of provisions, open daily 0900-1730. Sat 0900-1630 and Sun 1030-1300. (The Post Office is closed Weds pm, Sat pm and all day Sun). Should you need any washing done, the Island Laundry which is located near the estate office can do a same day service if your laundry is delivered to them before 1200.

At the brow of the hill beyond the Tresco Store there is a public telephone; continue along the road and a brisk 5-10 minute walk brings you to Old Grimsby, passing St Nicholas's C of E Church en route. The 'Island Hotel' which overlooks the northern side of Old Grimsby Harbour is open to non-residents for breakfast, lunch, dinner and bar. Showers are also available, along with freshly baked bread, and wine/spirits off-sales.

BRYHER

Bryher is the smallest of the inhabited islands and has a sleepy charm all of its own, much enhanced by the interesting topography which gives the island its name, being Celtic for "place of hills". Fishing, flower growing and, increasingly the tourist trade are the mainstay for the small population, but with limited accomodation ashore, it remains one of the least visited of the islands.

Like Tresco it is an island of contrast. Rushy Bay on the southern tip is a delightful white sandy beach with good bathing, the coastal footpath along the western side of the island provides a panoramic if somewhat chilling view of the Norrard rocks stretching away to the west, while a brisk foray over the tight turf of Shipman Down provides a spectacular view of the aptly named Hell bay, and a wild rock strewn path out towards Shipman Head. Watch Hill, one of the highest points in Scilly, was formerly used as a lookout for shipwrecks and has some of the best views in the islands.

Green Bay anchorage

From the shoal draught yachtsman's point of view Bryher is best known for Green Bay, an attractive and popular drying anchorage with excellent shelter from SSW through to north.

The easiest approach is immediately to the north of Merrick Island and south of Halftide bar, a higher bank which dries 2.8m LAT. Keep up more towards the northern side of the bay and sound your way into a suitable spot where you will dry out on a mostly flat sandy bottom. Take care to avoid the Brow Ledge and Three Brothers rocks which lie closer inshore on the south western corner of the bay.

Facilities

The promise of the grandly named settlement known as The Town belies the reality, for this is little more than a hamlet where you will find the 'Vine Cafe' and a phone box. However, although limited, the facilities on Bryher are surprisingly good, not least the Bryher Stores and Post Office (open 0900-1800 daily and 1000-1500 Sundays in season) renowned for Mrs Bushell's superb home baked pies, bread and pasties, and no cruise to Scilly is complete without sampling them! The international code flags S.H.O.P make it easy to spot from the anchorage, you can land in Kitchen Porth, or on the beach further to the south but keep clear of the jetty on Bar Point which is in more or less constant use. This unassuming structure is probably the most famous jetty in the UK; its high speed construction was witnessed by thousands of TV viewers when it was built as one of the many frantic 'challenges' faced by Anneka Rice.

The 'Fraggle Rock Cafe' also serves evening meals and those in search of a pint can stroll over to the bar in the 'Hell Bay Hotel' (follow signpost) on the western side of the island where meals are also available, either in the bar or full dinner.

There is no water or fuel available; limited chandlery, rigging and mechanical repairs are available from Blue Boats on the foreshore (Tel: 01720 423095).

SAMSON

Today barren Samson is the largest of the uninhabited islands, its only residents aggressive gulls and timid black rabbits. The burial cairns and ruined cottages are the only remaining evidence of the people who lived here, latterly in abject poverty, until the 1850s when Augustus Smith removed them to the larger islands and consigned Samson to cattle grazing and, briefly, a deer park, although this proved a failure when the deer discovered they could wade across at low water to enjoy the lusher delights of Tresco.

The best landing is on the fine sandy beach at Bar Point on the north-eastern corner, where it is possible to anchor in settled weather midway between Puffin Island and Bar Point in about 1m LAT, making a careful approach from Yellow Rock. This is very exposed at HW and is not somewhere I would recommend for more than a daytime stop.

ST MARTIN'S, TEAN SOUND and ST HELEN'S POOL

The north and north-western side of the Isles of Scilly are much less developed and less visited. St Martin's has a small population, both Tean, (pronounced Tee-Ann) and St Helen's are uninhabited, and for those seeking a quieter anchorage there are several possibilities, although the navigation is more demanding.

ST MARTIN'S

Surrounded by extensive shallows and drying sands along its southern side and exposed, rock encumbered bays along its northern coast, with the exception of Tean Sound, the long narrow island of St Martin's has few anchorages suitable for deeper draught boats. At Neaps fin keel boats of moderate draught will just be able to lie afloat in Higher Town Bay; for those able to dry out comfortably the choice is marginally better, although very dependent on the prevailing weather. Although the bays on the northern side of the island look enticing the numerous rocky ledges make this an area that should only be explored by the experienced in settled weather.

All of which is a shame, as the island is very attractive. With a backbone ridge of granite there is a distinct division along its length, the northern side is wild heathland edged with some deep cliff bound coves, in marked contrast to the fields and softer agricultural scenery of its gently sloping southern flank, which has an atmosphere in many ways reminiscent of rural Brittany.

From the holidaymaker's point of view the magnificent sandy beaches are the main attraction. The sailor knows the island best for the splendid rocket shaped red and

Tean Sound looking SW from St Martin's to Tean with Tresco in the distance. The yachts on the left are on the Hotel visitors' buoys. Just right of the moored fishing boats the slick on the water shows the position of Thongyore Ledge

white striped daymark which stands 56m above sea level at the summit of Chapel Down, the north-eastern corner of the island. This was erected in 1687 by Thomas Ekin (the date inscribed on the tower is incorrect) and it is well worth walking up to this vantage point for the views.

Approach and anchorage, Tean Sound

Tean Sound is a narrow but mostly deep passage between the western end of St Martin's and the deserted neighbouring island of Tean. It can be entered from seaward at any state of the tide but the approaches require considerable care as they are fringed with rocky ledges, and I would only recommend it in fair weather, good visibility and with an absence of swell. If there is a heavy ground sea running from the west or north-west seas can break heavily across the approach, particularly between Pednbrose and Pernagie, making it potentially very dangerous.

Ideally, try to time your arrival fairly soon after LW when most of the off-lying hazards will be visible, notably Black Rock (6m) and its associated ledges which extend northwards to Deep Ledges (least depth 0.6m LAT) and southwards to South Ledge (dries 1.4m LAT). Together they form a line of reefs nearly ½ mile in length, but it is possible to enter either east or west of this hazard depending on the direction of your approach.

From the east, give all of the northern shore of St Martin's and **White Island** a reasonable berth, until you have **Lion Rock** (8m) abeam to port. Keep a good 100m off this and its associated rocks and gradually ease round onto a heading of 180°20′T. This course can be confirmed by keeping the tallest (TV) tower on distant skyline of St Mary's just open of Goat's Point, the westernmost point on St Martin's.

Once Pernagie Island (9m) is on your port quarter you should edge more to port towards Plumb Island (13m) and its cluster of smaller islets to make sure you avoid Rough Ledge (dries 1.4m LAT). This lies almost midway between Plumb Island and Pednbrose (12m) the prominent island immediately to the north of Tean, which forms the western side of Tean Sound. Keep about 100m off Tinkler's Point and about 50m off Goat's Point on St Martin's to avoid Thongyore Ledge, a nasty rocky shoal (dries 1.4m LAT) on the starboard side of the channel.

Approaching from the west or north, keep up towards Round Island, taking care to avoid the Eastward Ledge which dries 2.9m a cable to the NNE of the island.

From here identify **Babs Carn**, a rocky bluff on the west side of St Martin's which is easy to spot, and the small flat island of **Pednbean** (1.8m) which lies to the east of Pednbrose. This is more difficult to identify from afar, but Pednbean in transit with Babs Carn on 154°T gives you the best approach line, leading midway between Pednbrose and the Corner Rock (least depth 0.3m) Once you have Pednbrose abeam, steer more to port towards Tinkler's Point, and the centre of Tean Sound.

The deepest water in the anchorage is taken up with four moorings belonging to the St Martin's Hotel, call sign *Santa Marta* VHF channel 16 or 12. Visitors are welcome to use these free of charge if they make use of the hotel facilities – lunch, evening meals and bar are all available to non-residents, as well as showers. Otherwise the charge is £7.50 a night.

Alternatively, anchor in the centre of the channel to the north or south of the moorings depending on the wind direction, but keep well clear of the end of the quay which is in regular use. Depending on your position, shelter can be found in winds from most directions except north, and it can be exposed in southerly winds at HW. At Springs the tide runs at up to 2 knots in the narrows which can create uncomfortable conditions with wind against tide, and the risk of dragging as the bottom is mostly rocky. A two anchor moor is definitely recommended.

In east or southerly winds an alternative anchorage can be found off Porth Seal, between Plumb Island and Tinkler's Point. There are sandy patches but also boulders and heavy kelp closer inshore and it is advisable to use a trip line.

Although the south approach to Tean Sound lacks any good transits it is easier than it looks on the chart and, once you have had a bit of practice, a good exercise in eyeball navigation provided you have reasonable light.

Make the approach at about half flood and steer 005°T from Crow Rock beacon until Broad Ledge (dries between 4.3m LAT and 5.3m LAT) is on your starboard beam. From here leave West Broad Ledge (dries 2.5m LAT but beware 0.7m LAT outlier on eastern side) to port and John Martin's Ledge (dries 3.9m LAT) to starboard then steer up for the moorings in the centre of Tean Sound. Even if covered, both rocky ledges will be easy to spot from the weed over them.

Higher Town Bay

Higher Town bay is sheltered from west through north to north-east, and although there are a few groups of rocks, with due care it is relatively easy to approach from Crow Sound, ideally after half flood. If you have sufficient water over Crow Bar you can use the useful stern transit, keeping the Crow beacon in line with the middle of the distant jagged backbone of the Haycocks rocks on the northern end of Annet, which should give you a heading of about 048°T. This will take you well clear to the south of the distinctive Guther's Island with its twin flat topped rocks and its associated rocky outliers, leaving the three Damasinnas

Higher Town Bay looking west towards the quay

(known locally as the Sinners) rocks (drying between 2.4m and 1.5m LAT) on your starboard hand and the Three Rocks Ledges (drying between 0.9m and 0.7m LAT) to port.

Here again, a good lookout on the foredeck will be able to spot both these hazards from the large amount of weed clinging to them. As you draw abeam of Cruther's Point and the small landing quay, wait until you have the two boathouses on the dunes behind the beach open of the end of the quay before turning inshore – this will clear the last remaining hazard, isolated **Harry's Rock** (dries 1.2m LAT), a cable south-east of the quay. Anchor anywhere in the western end of the bay, but keep clear of the approach to the quay which is used by the ferries. Do not leave dinghies tied to the quay, but land anywhere on the beach.

Facilities

Do not be misled by the prospect of no less than three Towns, for here again these are little more than hamlets, Higher Town in the centre, Lower Town at the western end and Middle Town in between. Lower Town has a handy fruit and veg shop 'Lindy's Locker', with a limited selection of other provisions, open daily 0930-1730. Just above it on the hillside you will find the 'Seven Stones' pub, which does bar meals and is open for evening meals 1930-2100.

Middle Town has little more to offer than a public phone box; Higher Town Post Office Stores and off-licence (open 0900-1700 daily, closed for lunch, and open for an hour on Sunday mornings) has a reasonable selection of tinned food and frozen produce, gifts and books but no fresh food. There is also a public phone, tea room and crafts/gift shop. There is no fuel available on the island. Water in small quantities (cans) can sometimes be obtained from the St Martin's Hotel.

ST HELEN'S POOL

St Helen's Pool is an open roadstead fringed by Tean, St Helen's Island and Tresco. At one time this was the favoured bolt hole for larger vessels when St Mary's Roads became uncomfortable. On St Helen's, there is a legacy from the days of sail; the small ruined building, dating from 1756 and known as the pest house, was once an isolation hospital for disease ridden seamen. St Elidius' hermitage, the oldest Christian building in Scilly, and a small complex of excavated church building dating from the 8th-12th century will be found nearby. Behind them a steep path leads to the heathery summit of the island from where there are spectacular views of St Helen's Pool, the lighthouse on Round Island and Men-a-vaur.

Tean was inhabited until the latter part of the 18th century and was a centre for burning kelp to make fertiliser. The remains of the few small cottages can still be found on the southern foreshore; remember though, that the island is voluntarily closed between 15th April and 20th July when the ringed plovers are nesting.

The total lack of facilities and sheer isolation of St Helen's Pool are its main attractions and it is often a much less crowded alternative to the more popular anchorages. The shelter is excellent at Low Water, when it is virtually landlocked. At High Water it can be exposed for a time from the north-east, north-west and south-east and depending on your position if there is any prevailing swell it will be felt. There is however, plenty of depth, usually plenty of room and the holding is good. Although you will certainly feel fully exposed to the force of the wind in a blow, it will not generate a great deal of sea and this, in many ways, is as safe as any anchorage in Scilly in such conditions. For boats able to dry out comfortably, **West** and **East Porth** on the southern side of Tean provide excellent shelter.

St Helen's Pool looking south over the East and West gap rocks to Tean where the distant yachts are moored in Tean Sound. Hedge Island is the distinctive lump in the centre of the picture with St Mary's beyond

Isles of Scilly

Approaches and anchorage.

St Helen's Pool can be approached in several differing ways – from Crow Sound, which is shallow, but no problem with sufficient rise of tide. From Old Grimsby, again shallow but easy with a rising tide; from St Mary's Sound towards HW or from seaward from the north through the passage known as St Helen's Gap. This is deep enough in itself but the inner shallow bank between it and the Pool means that here again you will need sufficient rise of tide.

The tide runs strongly across this approach, it is both narrow (about 300m wide) and flanked by rocky ledges, leaving little room for mistakes. Personally I would recommend one of the 'overland' approaches for a first visit, you can then reconnoitre St Helen's Gap for future reference, or alternatively leave that way.

Approaching from the vicinity of the 'Hats' buoy in Crow Sound wait until half flood. Steer 322°T , which will take you past the distinctive **Guther's Island**, and, more important, Higher Ledge (dries 4.0m LAT) which should just still be showing. In the far distance ahead, to the left of the rounded island of St Helen's the unmistakeable pointed summits of **Men-a-vaur** are the mark you now need – get the Landing Carn, a prominent lump of rock on the western end of St Helen's, in line with the gap between the two highest summits on Men-a-vaur and this 322°T transit will take you safely into the anchorage, passing between the Chinks Rocks (awash LAT) and Hunters Lump (dries 0.9m LAT). This is the narrowest part of the approach, proceed slowly with a lookout on the bow.

St Helen's Pool leading marks: the rocky landing carn centred on Men-a-vaur's distinctive peaks

Once the impressive lump of **Hedge Island** is abeam you are entering the Pool, which extends for nearly mile, with depths varying between 2.5m LAT and 7.0m LAT. If you intend to anchor close to St Helen's Island beware the sandy spit (dries 2.9m LAT) which hooks well south and west from the island, and sound in until the pest house ruin on the shore is bearing about north-east.

At low water, when the extensive reefs of Golden Ball Brow are uncovered to the north-west they form a perfect natural breakwater and the shelter is excellent. At the top of the tide however, if there is any swell, you are likely to feel it here, and it will pay to look around for a more comfortable berth further to the south, either east of Foreman's Island, or west of Old Man, depending on the wind direction. Throughout the Pool, the tide follows much the same pattern as in Old Grimsby, running south-east at up to

2 knots Springs for nearly eight hours, beginning HW Dover -0040 then turning north-west for four hours.

The approach from Old Grimsby is a simple bit of eyeball navigation, ideally from half flood onwards. It is easiest if you pass south of Lump of Clay Ledge (dries 1.4m) and leave Little Cheese Rock several boat lengths to port before heading up towards St Helen's. Little Cheese (0.7m) is visible even at HW, when covered the surrounding ledges are easy to spot from the weed on them.

The approach from St Mary's Sound is more demanding, and uses the same marks described in the southern exit from Old Grimsby – St Mary's Telegraph tower in transit with Crow Rock Beacon, 162°T. On this the Cones Rock (dries 0.6m LAT) and Diamond Ledge will be left close to port so keep a good lookout, and you will then pass midway between Little Pentle Ledge (dries 2.8m LAT) and West Craggyellis Ledge (dries 0.8m LAT). Both will be covered, but Great Pentle Rock (1.7m) at the western end of the Pentle Ledge gives you a good indication of when you have passed these hazards.

Once Lizard Point on Tresco is abeam keep slightly to port to avoid Tea Ledge (dries 3.7m LAT) which should be visible from the weed on it, and then ease over to starboard, leaving Little Cheese Rock on your port hand and thence straight up into the anchorage.

If you climb to the summit of St Helen's Island to take in the view which is definitely best at low water, you you will also be able to see clearly **St Helen's Gap**, the narrow (100m) passage out to the north between the West (0.9m) and East (2.3m) gap rocks. Although both are visible at HW, ledges extend beyond them for some distance particularly to the north of the East Gap Rock.

Approaching from seaward, keep up to Round Island, leaving it 2-300m to starboard and then steer to leave Didleys Point, the eastern end of St Helen's, about 100m to starboard. The Gap Rocks should be clearly visible from here, steer for the West Gap (starboard hand) Rock until you are almost abeam of the East Gap Rock when you should ease more to port, passing midway between the two. Keep this course to the SSW and do not be tempted to haul round too rapidly into the Pool or you will fall foul of the sand spit (dries 2.9m LAT) which extends a good 200m from the island shore.

THE EASTERN ISLES

With calm weather and good visibility, Admiralty chart 883 and considerable care, for the more adventurous the Eastern Isles can make an interesting foray, with a number of temporary daytime anchorages, depending on the prevailing conditions, and the chance of seeing a few seals along the way, particularly in the vicinity of Menawethan. It is, however, not an area I would recommend for an overnight stop except in exceptionally settled weather.

The easiest approach is south of Biggal and to the east of Ragged Island, keeping over to the west side of Great Ganilly. There are several entrance graves on the Arthurs but **Nornour** is probably the most interesting of these now deserted islands for here, along the southern edge of the island, you will find the impressive excavated Bronze Age settlement for which the island is famous.

Isles of Scilly
ALL-TIDE ANCHORAGES AT A GLANCE

(Note: this guide is only intended as a general indication of the best likely shelter in winds from the given direction – the overall comfort of the anchorage will be greatly affected by the strength of the wind and the prevailing swell and tide.)

Wind	Shelter options	Comment
North:	St Mary's Pool	Good if closer inshore
	Porth Cressa	Very good in absence of swell
	New Grimsby	Good, but prone to swell
	Old Grimsby	Good
	St Helen's Pool	Good in lee of St Helen's
	The Cove	Very good, except when bar covered at HW

Wind	Shelter options	Comment
NE:	St Mary's Pool	Very good
	Porth Cressa	Very good
	New Grimsby	Very good
	Old Grimsby	Good
	Porth Conger	Good, if well in
	The Cove	Good
	St Helen's Pool	Good in lee of Old Man
	Tean Sound	Good at south end

With Porth Cressa Brow uncovered, the bay is a very different picture at LW. Note the Wras and Biggal beyond

Isles of Scilly

Watermill Cove at Low Water showing how far the reef extends to seaward from Block House Point

Wind	Shelter options	Comment
East:	St Mary's Pool	Very good
	Porth Cressa	Very good
	Porth Conger	Very good
	The Cove	Good if well in, but possible surge
	New Grimsby	Very good.
	St Helen's Pool	Good in lee of Tean
	Tean Sound	Very good.
SE:	St Mary's Pool	Very good
	New Grimsby	Good
	Porth Conger	Very good, except at HW if bar covered
	Tean Sound	Good at northern end
South:	St Mary's Pool	Very good
	Watermill Cove	Very good, but possible swell around HW
	Porth Conger	Very good except at HW if bar covered
	Old Grimsby	Good
	New Grimsby	Good, except towards HW, with wind against tide
	St Helen's Pool	Good

Wind	Shelter options	Comment
SW:	New Grimsby	Good, particularly at north end
	Old Grimsby	Very good
	Watermill Cove	Very good, but possible swell at HW unless well in
	Porth Conger	Very good
	The Cove	Good, if well in, but subject to swell
	St Helens's Pool	Good
West:	New Grimsby	Very good
	Old Grimsby	Very good
	Watermill Cove	Very good, if well in
	Porth Cressa	Good, if well in, but subject to swell at HW
	Porth Conger	Good, if well in
	The Cove	Very good
	Tean Sound	Good
	St Helen's Pool	Good in lee of Norwethal
NW:	Porth Cressa	Very good
	The Cove	Very good
	Old Grimsby	Good, if well in
	St Helen's Pool	Good, but swell likely around HW. Best toward south end of pool

Passages
LAND'S END TO PENTIRE POINT

Favourable tidal streams
Land's End Bound west/north: 1 hour after HW Dover
Bound south/east: 5 hours before HW Dover

Passages—Land's End to Pentire Point
Charts for this section of coast are:
BA: 2565 St Agnes Head to Dodman Point. 1148 Isles of
Scilly to Land's End. 1149 Pendeen to Trevose Head. 1156
Trevose Head to Hartland Point. 1168 Harbours on the North
Coast of Cornwall is particularly useful
Imray: C7 Lizard Point to Trevose Head. C58 Trevose Head
to Bull Point
Stanford: 13 Start Point to Padstow
French: 2218 Du Cap Lizard à Trevose Head

Beacon	Range (M)	Frequency (kHz)	Call Sign
Marine beacon			
Round Island	150	298.5	RR
Coastal Radio			
Land's End Radio, working channels 27, 88, 85, 64			

As a cruising ground there is, sadly, little to recommend the north Cornish coast. In the prevailing southwesterly winds it presents an exposed and rugged lee shore with just a few drying harbours that cannot be guaranteed as places of refuge – an unwholesome combination and a natural deterrent for pleasure boating. Bound round Land's End, and up the Irish Sea your course will soon take you well away from the coast, an offing that you should endeavour to maintain.

Although there is little real cruising potential on this dangerous stretch of coast, it has become increasingly used by yachts on passage to and from South Wales and the Bristol Channel to the Isles of Scilly and the West Country during the summer months. Most would, I'm sure, agree that it is an area they try to pass through as quickly as possible, waiting for the right conditions, and completing it in one leg. However, there will obviously be occasions when circumstances dictate a need to put in somewhere, or, just possibly, in very settled weather and an offshore wind, a little bit of cruising *can* be contemplated, and it is for this reason that this shorter passage section is included.

Land's End, the most westerly point in England, is not a particularly distinctive headland, being somewhat lost against the rest of this noble stretch of pink/grey granite

Land's End

157

Seen from the inshore passage at half-tide the Longships reef is a sombre sight. Kettle's Bottom rock is on the right

cliffs which rise to over 70m along one of the most rugged sections of coast in southern England. The large white hotel building is part of the privately owned Land's End tourist complex, and further to the north the famous 'First and Last House' in England is high on the cliff top.

A mile offshore, the infamous **Longships** reef has claimed many ships and the tidal streams are strong and unpredictable in their vicinity. Although the north going flood begins one hour after HW Dover, it does not turn north-east along the north Cornish coast until two hours later attaining over 2 knots at Springs. Five hours before HW Dover, the ebb begins to run south-west back along the coast.

There is a convenient inshore passage between the Longships (Iso WR 10s W19M R18/15M) and the mainland but this should only be attempted *in settled conditions and good visibility and never at night*. Pass a quarter of a mile to seaward of Land's End, from where the **Brisons**, two conspicuous rocky islands to the north, provide the best transit. The highest point of the highest island (27m) should be kept just open to the west of the highest point of the low island (22m), a bearing of 001°T. **Kettle's Bottom,** forming the inshore, easternmost, extremity of the Longships rocks dries 5.2m LAT, and **Shark's Fin**, the most northerly reef dries 3.2m LAT. If passing to the west of the Longships, give the whole area a good berth, and if bound up the coast maintain a course well north of east for just over a mile to clear the Shark's Fin. At night, the north-eastern red sector of the Longships covers all hazards along this section of coast and should not be entered until Pendeen lighthouse (Fl (4) 15s 27M) opens. With jagged rocks and breaking water to seaward, and the tall, lonely finger of the lighthouse with its precarious helicopter

landing pad, this is a menacing stretch of water where you will probably feel little inclined to linger.

The large sandy sweep of Whitesand Bay forms a break in the cliffs, and the village and tiny harbour of **Sennen Cove** will be seen at its southern end with its lifeboat house, the most south-westerly station in England. It was here, in 1794/95 that Samuel Wyatt the Trinity House architect assembled, numbered and shipped out the granite blocks to Carn Bras rock to build the first Longships lighthouse, replaced in 1873 with the present structure by Sir James Douglass. The keepers were often stranded here during winter storms and in 1966 it was one of the first lighthouses to receive much needed supplies by helicopter, the lantern house windows padded out with mattresses in case the rotors touched!

There is a fair weather anchorage in offshore winds just off Sennen village, inside the Cowloe Rocks, but this is very exposed and not recommended. Passing to the north of the Brisons the tidal streams run strongly past **Cape Cornwall**, a distinctive, cone shaped headland, uniquely, the only cape in England. Topped by a conspicuous ruined chimney this is one of the first indications of the extensive mining operations that once covered this stretch of coast. For the next ten miles, the cliffs are dotted with old pumping houses, chimneys and other buildings, from which the underground workings extended far beneath the sea bed, in places well over a mile out into the Atlantic.

The Vyneck is an isolated rock three cables north-west of Cape Cornwall – from here on a course a mile to seaward will clear all hazards except the overfalls which extend westwards from **Pendeen Head**, a bold headland. On it, the squat white lighthouse of Pendeen Watch (Fl (4) 15s 27M) looks out over the Wra, or Three Stone Oar, a

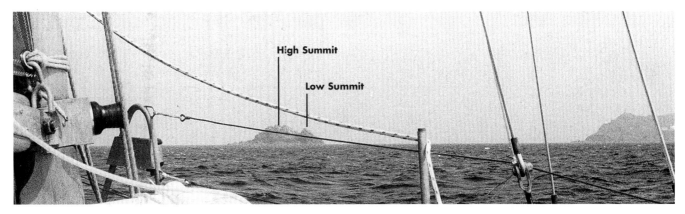

Longships inshore passage marks: high summit just open of low summit on the Brisons. Cape Cornwall is on right

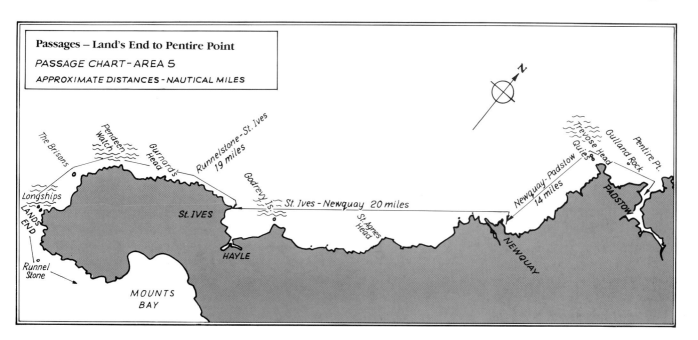

Passages – Land's End to Pentire Point

PASSAGE CHART – AREA 5

APPROXIMATE DISTANCES – NAUTICAL MILES

group of small rocky islands, just under a half mile off-shore. The coast follows a north-easterly direction from here on, a continuous unbroken line of impenetrable granite cliffs and one of the major sea cliff climbing centres in England. Along it, and in the approaches to **St Ives Bay,** a good lookout should be kept for pot and net buoys; in spite of the exposure of this coast it is much worked by local boats.

This four mile wide bay is backed by the sandy beach and dunes of Hayle Towans, and guarding its far northern end the lighthouse (Fl WR 10s W12M R9M) on **Godrevy Island** is close inshore. Built in 1859, this was always a controversial siting for it gives little indication of the notorious reef, **The Stones,** which extends nearly a mile to the north-west of the island. A lightship or lighthouse on the outermost rock was the favoured option but was turned down because of the cost. An area of strong currents, and drying rocks, many awash at HW, their outer limit is marked today by the 'Stones' BY north cardinal buoy (Q), and the area is covered by the red sector of the Godrevy light 101°-145°T. The Sound, the inshore passage between Godrevy and the Stones is half a mile wide and not recommended without local knowledge; there are overfalls, particularly with wind against tide, and the additional hazard of numerous pot buoys – this is no place to be caught with a fouled propeller if motoring. Pass well to seaward of the whole area.

Bound north from St Ives Bay the flood begins about 2 hours after HW Dover, (2 hours after local LW). **Bound south,** ideally leave around local HW, which will entail pushing the tide for the first hour, but will ensure a fair tide round Land's End and into Mount's Bay.

From the Stones buoy it is just over 17 miles to the next possible drying harbour at Newquay, along an impressive but unwelcoming stretch of coast, mostly high, crumbling cliffs, averaging between 40 and 75 metres in height. Along them there are numerous sandy coves and bays such as **Porthtowan** and **Perranporth**, holiday resorts, particularly popular with surfers, the almost perpetual ground

swell producing ideal conditions for the sport. Numerous small rocky islets lie close inshore, but with the exception of Bawden Rocks, two small islands a mile north of **St Agnes' Head** there are no off-lying dangers. Unless heading for Newquay, the direct course to **Trevose Head**, just over 22 miles to the north-east, takes you safely two to three miles offshore.

Godrevy Lighthouse

Trevose is a bold headland with a prominent lighthouse

Inland, higher ground runs parallel to the coast, and a very conspicuous feature is the large obelisk on the skyline south-east of **Portreath.**

Trevose Head lighthouse (Fl R 5s 25M) is a prominent white tower on a precipitous headland with two large rocky islets, the **Bull and Quies**, nearly a mile to seaward. There is an inshore passage between them and the coast but this should not be used without local knowledge, and here the tide runs hard. Give the islets and the headland a wide berth. From here **Padstow Bay** begins to open with **Pentire Point** forming its distant northern extremity just over five miles away.

With no possible shelter in winds between south-west and north-east throughout this passage area if the weather deteriorates and a blow looks likely, particularly from the south-west, the advice in the Admiralty Pilot *'to seek a good offing'* is probably as sensible as you will get. Padstow is the only place where complete shelter will be found once you are inside the harbour. However, this is tidal, and only accessible 2 hours either side of HW, and the Camel estuary on which it lies is approached over a bar that becomes very dangerous in strong onshore weather. The decision to run for Padstow in deteriorating conditions should be considered very carefully, with particular regard to the state of the tide, aiming to arrive between half-flood and HW. With any ground swell, once the ebb commences, the shallow waters of Padstow Bay can become very hazardous, and unapproachable in strong winds and sea from the north-west.

In fog or poor visibility additional aids to navigation along this section of coast are the Longships (Horn 15s), Pendeen (Siren 20s), Stones Buoy (Whistle) and Trevose Head (Horn (2) 30s).

ST IVES

Tides: HW Dover –0605
Charts: BA: 1168. Stanford: 13. Imray: C7
Hazards: Harbour dries, exposed in onshore winds. Hoe Rock, and conical green buoy to NE (both unlit),
          Carracks rocks to SE (unlit)

From a distance St Ives Head looks like an island and entering the bay this should be given a good berth to avoid Hoe and Merran rocks on its north-eastern side, which both dry. A conical green buoy marks the end of the ruined outer pier and should be left to starboard. Smeaton Pier, the eastern arm, has an old white lighthouse at its end, at night this is marked by (2FG vert) and the west pier (2FR vert), enclosing a harbour which dries completely at LAT, but is normally accessible after half flood. It has a firm sandy bottom, and there is a fleet of quite large fishing and trip boats based here during the summer months which lie on heavy fore and aft drying moorings or alongside the quay. Shelter is good in south and westerly weather, but anything further north can send a considerable sea into the bay and a heavy swell within the harbour.

In favourable conditions, anchor about 100m south-east of the harbour entrance in about 3m, but watch out for the large number of buoys marking the keep pots towards Porthminster beach. Deep keel boats can only really berth

inside the outer end of Smeaton Pier, where they will dry, but this is not the most comfortable place as there is often a surge, and there is also plenty of commercial activity. However, bilge keelers or boats able to take the bottom are well catered for, with eight drying visitors' moorings in the harbour just off the prominent 'Woolworths' store on the waterfront. Charges are £7.75 per night inc VAT.

During the season the town reels under the assault of the tourists, its picturesque narrow streets and alleyways, harbour and wide sandy Porthmeor Beach on the seaward side of St Ives Head, providing all the essentials of a seaside holiday. In spite of a certain amount of inevitable commercialism, the town retains much of its unique atmosphere, a factor that has made it a popular haunt of artists for many years, even more so with the arrival of the Tate Art Gallery's western outpost which overlooks Porthmeor.

Eric Ward, the Harbour Master, will help with any problems. His office (Tel: 01736 795018) does have a VHF but no constant watch is kept. There is fresh water on the piers, Calor and Camping Gaz from the Fishermen's Co-operative, but no easily obtainable fuel. All other normal requirements can be found, banks, Post Office and provisions, there are many cafes, pubs and restaurants, and a very handy toilet and public shower block in the car park just behind the harbour front. There are road and main line rail connections. A lifeboat is stationed at St Ives, launched on a trailer across the beach.

St Ives Bay stretches away to the north-east backed by the extensive sand dunes and beach of Hayle Towans. Hayle, in the south-eastern corner is a run-down, tidal harbour used by a few local fishing boats. Approached over a dangerous bar and tricky entrance channel through a large expanse of drying sands, it is not recommended for visitors.

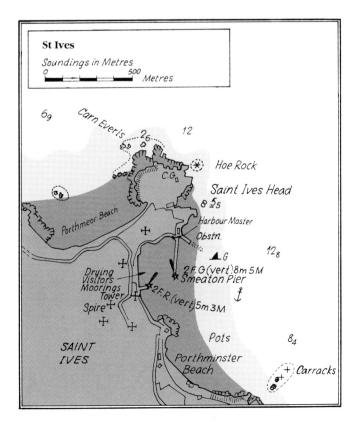

NEWQUAY

Tides: HW Dover –0604
Charts: BA: 1168. Stanford: 13. Imray: C7
Hazards: Small harbour dries beyond entrance. Heavy surge in onshore winds and dangerous to approach. Old Dane and Listrey
rock to N (both unlit)

In fair weather and, ideally, offshore winds Newquay is a pleasant small drying harbour that is feasible for an overnight stop. Midway between it and Godrevy, St Agnes' Head is prominent, 91m high, with steep cliffs, and inland the isolated St Agnes' Hill (200m) covered in heather and gorse is surmounted by a beacon. This was another site of intensive tin mining activity in the 1800s, and several prominent ruins can be seen along the shore. Just north of St Agnes' Head, Trevaunance Cove was but one small port exporting tin and copper frequently rebuilt between 1700 and 1920 as gales swept away the massive granite walls.

Closing the land from the south, East and West Pentire Points form the entrance of the Gannel, a silted-up river mouth, which forms a fine sandy beach, and, Fistral Bay, renowned for surfing, leads up to **Towan Head,** the entrance to Newquay Bay, which has the large 'Atlantic Hotel' prominent on its highest point.

Although there are no off-lying dangers, overfalls occur up to half a mile off Towan Head, and it is wise to give it a reasonable berth before turning into the Bay, when the

eastern extremities of Newquay's extensive hotels and houses will come into view along the skyline right down to the cliff-backed beaches.

The harbour lies in the south-eastern corner, hidden behind a headland, ENE of which the Listrey Rocks, least depth 0.5m LAT and 1.2m LAT, lie 300m offshore. Keep well out into the bay before heading in towards the harbour walls as they come into view. The whole of the inner part of the bay dries at LWS and the harbour is normally only accessible for average draught boats 3 hours either side of local HW (HW Dover -0604). If waiting for the tide, anchor off in the bay, which is well sheltered from the south and west.

Visitors normally lie alongside the inner end of the south quay, where there is a clean, hard sandy bottom, and you will dry clear of the local fishing boats which lie on heavy fore and aft moorings in the middle of the harbour, which suffers from a surge if there is much ground sea running. The Harbour Master is very helpful and is usually around 0830-1730 in the summer and his office is on the south pier, (Tel: 01637 872809). Charges are 25p per foot per night, and water and diesel are available from him on request. There are public toilets close by, and also the licensed clubhouse of the Newquay Rowing Club, where temporary membership is available, which saves the trek up into town for a pint. These dedicated enthusiasts own several of the beautifully restored Scilly pilot gigs including *Newquay*, the oldest one still afloat, built in 1812.

One of the largest holiday resorts in Cornwall and the 'surfing capital of Britain', the busy main centre of Newquay is a short walk from the peaceful harbour. All normal requirements are available: Post Office, banks and provisions, launderettes, pubs, cafes and restaurants with road, rail and air connections.

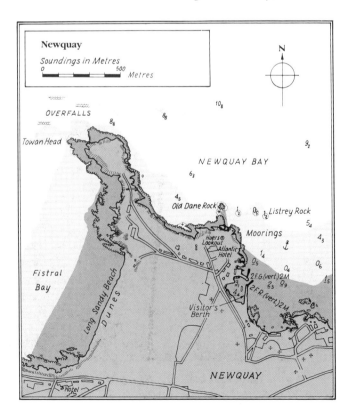

PADSTOW

Tides: HW Dover –0550. Range: MHWS 7.3M – MHWN 5.6m. MLWN 2.6m – MLWS 0.8m. Streams attain over 2 knots in channel off harbour entrance at Springs. Inner harbour (visitors) floating basin entered by tidal gate 2 hours either side of HW

Chart: BA: 1168. Stanford: 13. Imray: C7, C58

Waypoint: North Pier End 50°32'48N. 04°56'10W

Hazards: Gulland, Newland, Gurley, Chimney and Roscarrock Rocks (all unlit) and many pot buoys to seaward. Wreck west of Stepper Point (unlit). Doom Bar and much of river and outer harbour dries. Buoyed channel (lit) but liable to change. Approach can be dangerous in onshore wind and sea. Busy tidal fishing port

It is another thirteen miles from Newquay to the mouth of the River Camel, and Padstow lies two miles upstream. This is a most attractive estuary, a spectacular area of golden drying sands, with a shoreline of low cliffs, fine beaches and sand dunes providing some lovely walks. However, the mouth of the river is restricted by the Doom Bar, a large sandbank drying at low water. During strong onshore winds, or a heavy ground sea this lives up to its melancholy name particularly around low water when the sea can break right across the entrance – conditions that are not always obvious from seaward. With care, though, the River Camel is not difficult to enter in daylight and yachts of normal draught will have ample water from half-flood onwards. The approach at night is straightforward once you have picked up Stepper Point light, which only

has a 4M range. Now that it is possible to lie afloat in the inner harbour it has become increasingly visited by pleasure craft, and the ongoing improvement to facilities reflects this.

The approach from the north or south is unmistakable as Trevose Head, four miles to the west, is a steep headland, 80m high, which rises higher than the surrounding hinterland, giving it the appearance of being a separate island from a distance. In addition, the **Bull** and **Quies**, prominent large rocks, lie a mile further to the west. The lighthouse is on the north-western corner of the headland and dates from 1847; Padstow lifeboat house is on the north-eastern side of the headland, re-sited here after the original station in the river fell prey to the encroachment of the Doom Bar, which, according to local legend was

The mouth of the Camel seen from St Saviour's Point at half ebb. The breaking seas clearly indicate the position of the Doom Bar. Stepper Point is on the left, Newland is in the centre and Pentire Point is on the right

created by a mermaid shot by a local man who thought she was a fish. Cursing him with her dying breath, she threw up a handful of sand which turned into the Bar, vowing that *'henceforth the harbour should be desolate!'*

Tidal streams run strongly off Trevose; over two knots at Springs, particularly between the rocks and the headland and you should keep to seaward of the Quies before heading up into Padstow Bay, passing inside **Gulland Rock**, a prominent rocky island 28m high. There are several isolated hazards further inshore; Gurley Rock, least depth 3m LAT, and Chimney Rock, 2. 3m LAT, not normally a problem to boats of average draught but worth remembering near low water, particularly if there is any ground swell running. **Stepper Point,** a bold and rounded grassy headland, footed with cliffs, forms the south side of the river mouth, and is easily identified by the large stone daymark, like a truncated factory chimney. There is a lookout at its eastern extremity, and a light on an iron pillar (L Fl 10s 4M). Three cables WNW of the daymark there is a dangerous wreck, almost awash at LW, so give the headland a wide berth and head up towards **Pentire Point**, the north-eastern arm of the bay, another bold headland, with the distinctive island of **Newland**, a pyramid 35m high, half a mile to the north-west. Approaching from the north, it is best to keep well to seaward of Newland, as Roscarrock, a rock with least depth of 0.8m LAT, lurks three cables west of Rumps Point, the north-eastern corner of Pentire.

Enter the river on a flood tide, ideally no earlier than 3 hours before local HW (HW Dover -0550) and do not attempt it in any ground swell from the north-west, or if breaking water can be seen. The **Doom Bar**, drying sand, fills the south-western corner of the river mouth, and opposite, the houses of Trebetherick sprawl along the low cliffs. These are fringed with rocky ledges, and the channel between them and the sands runs due south, the western end of Pentire Point, providing a useful back bearing. Depths reduce quickly from 3m to a least depth of 0.8m LAT two cables north of the first channel buoy 'Greenaway', a red port hand can (Fl (2) R l0s), and 'Doom Bar', two cables further south, is a conical green starboard hand buoy (Fl G 5s). At low water the channel is a cable wide and depths vary between 2m and 0.4m LAT as far as St Saviour's Point, known locally as 'Ship-me-pumps' a

Outer harbour entrance showing banks and channel

quarter of a mile downstream of the harbour. There is a prominent monument on this rocky headland and a red and green middle ground buoy (Fl R 5s) abeam of it where the channel divides. The main channel to the harbour lies to starboard of the buoy, while the river bears away towards the village of Rock on the eastern shore.

Just south of the buoy, there is 'The Pool' with an average depth of 3m and a number of local moorings in it. It is possible to anchor clear of them, if waiting for sufficient water to get into the Inner harbour, although at Springs the tide can run hard, between 4 and 5 knots at times on both the flood and ebb.

A large area of drying sandbanks fills the centre of the river and the narrow approach channel to Padstow also dries almost completely at Springs. Hold to the western shore and leave the two green beacons with triangular topmarks close on your starboard hand, the first (L Fl G 10s) and the second (QG). Just short of the harbour entrance there is a port hand can (Fl R), on the western side if the sands, and the outer pier ends are marked with (2FR vert) and (2FG vert) port and starboard.

Proceed under power and watch out for fishing boats under way, particularly in the harbour mouth, where the main stream runs strongly at right angles to the entrance. South Dock, the commercial harbour, lies immediately to port, a long narrow, drying basin, but visitors should continue straight ahead through the tidal gate into the Inner Harbour which is open 2 hours either side of HW, day and

Padstow harbour approach. Middle ground buoy and St Saviour's Point

Inner harbour entrance showing tidal gate just opening

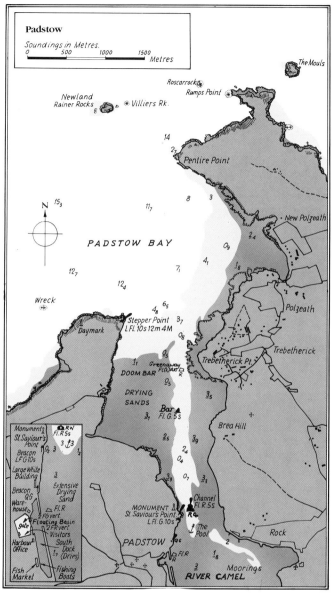

Padstow

Soundings in Metres.
0   500   1000   1500
                         Metres

Facilities

A small, unspoilt town, with attractive and colourful old buildings lining the quayside, although popular with holidaymakers Padstow has managed to avoid much of the overt commercialism of some of the other north Cornish resorts. Formerly a vital port for north Cornwall, exporting tin, copper, slate, granite, china clay and grain, and importing coal, salt and timber, it was also important as a shipbuilding centre. By the beginning of this century, however, commercial trade had dwindled steadily with the silting of the Doom Bar and the advent of road and rail transport and it would seem that the mermaid's wish had been fulfilled, for little more than a small fishing fleet remained. However, in recent years this has grown considerably, with a new fish co-operative, evidence of the resurgence in prosperity. Pot and net boats land large quantities of shellfish here, as a look in the large wet tanks on the Fish Quay will reveal, and large beam trawlers also work out of the port. Coasters occasionally discharge cargoes of fertiliser and roadstone and there is a sand dredger working in the river.

In the prominent red brick building on the north side of the harbour there are visitors' showers (£2 per boat, first night free) toilets and a laundry, along with a Tourist Information Centre. There are Midland, Lloyds and Barclays banks, provisions, including a 'Spar' open late and on Sundays, launderette, chandler and water, diesel and petrol are available through the Harbour Master. Not lacking in pleasant pubs, notably the 'Harbour Inn' and the 'Old Custom House', and a number of restaurants, the walks out towards Stepper Point and beyond are a delightful way to work up an appetite or thirst. It is unlikely that you will be in the area as early as May 1st when the famous 'Obby Oss' festivities take place when strange creatures dance through the streets to celebrate the coming of summer, but you might coincide with the annual regatta during August. There is a small museum in the centre of town, and the Shipwreck Museum on South Quay, where Padstow Sailing Club is also situated.

However, much of the sailing activity in the estuary is centred on Rock, on the opposite shore, where there are several deeper pools and a large number of local moorings, and some drying Harbour Commission moorings also available for visitors athough these tend to be hired on a weekly basis for people holidaying in the area. A much quieter little village, with many holiday homes, Rock is linked by a regular ferry to Padstow, and is home of the Rock Sailing Club which has a fine clubhouse, an old converted grain warehouse on the quay with excellent facilities. Visitors are welcome to use the bar and showers, and snack meals are also available. The sand dunes and clean beaches running to seaward from Rock towards Brea Hill are most attractive, and it is a particularly fine area for dinghy sailing.

The Camel, once a busy waterway, is still navigable on a good tide as far as Wadebridge, about five miles inland where yachts can dry out alongside Commissioner's Quay, but this definitely needs some local knowledge or reconaissance in the dinghy as the channel is tortuous and unmarked.

night, weekends and Bank Holidays. This hydraulic bottom hinged gate was installed during 1988/89 as part of a flood prevention scheme and now maintains a depth in the harbour of between 1m and 2.5m – 3m at neaps depending on your berth. More water can usually be retained to accommodate deeper draught boats if the Berthing Master is informed on arrival. Berth alongside the south quay where you will remain afloat in attractive surroundings. Charges for a 30 footer are £11.30 inc VAT for the first night – this includes a one off annual payment of river dues – thereafter the nightly charge is £6.70 inc VAT.

The Harbour Office is close by on the quay and is open 0800-1700 weekdays, and from 2 hours before to 2 hours after HW, on every tide. and you should try to report by VHF before arrival or as soon as you have berthed. Visitors are requested not to leave vessels unattended for any length of time without first consulting the Harbour Master.

The Harbour Commission sometimes have visitors' moorings available in the river for vessels up to 12m; either call *Padstow Harbour* on VHF channel 16, working channel 12, during office hours or telephone ahead for availability.

PORT GUIDE: PADSTOW
AREA TELEPHONE CODE: 01841

Harbour Master: Captain John Hinchliffe, Harbour Office, Padstow, PL28 8AQ (Tel: 532239) Mon-Fri, 0800-1700
VHF: Channel 16, works 12, call sign *Padstow Harbour*, office hours and 2 hours either side of HW
Mail Drop: c/o Harbour Office
Emergency Services: Lifeboat at Trevose Head. Falmouth Coastguard
Anchorages: 'The Pool' quarter mile downstream of Padstow . Drying, clear of moorings at Rock
Moorings/berthing: Inner harbour, afloat at all times , access through tidal gate 2 hours either side of local HW, day and night, weekends and Bank holidays. Harbour Commission moorings in river sometimes available on application
Charges: For 30 foot boat, first night £11.30 inc VAT and river dues, thereafter £6.70 per night inc VAT
Water Taxi: None, but regular ferry Padstow to Rock
Marina: None
Phones: On Quay
Doctor: (Tel: 532346)
Hospital: Treliske, Truro (Tel: 01872 74242)
Churches: All denominations
Local weather forecast: Harbour Office
Fuel: Harbour Master can arrange diesel and petrol
Gas: Calor/Gaz, Rigmarine, South Quay
Water: On quay
Tourist information office: Red brick building, north side harbour
Banks/cashpoints: Barclays and Lloyds have cashpoints. Midland, no cashpoint
Post Office: Duke Street
Rubbish: Bins on quay

Showers/toilets: In red brick building, north quay. Public toilets on quay. Toilets/showers, Rock Sailing Club
Launderette: In red brick building, north quay. Church Street
Provisions: Most normal requirements. EC Weds but many shops remain open. 'Spar' open till 2130, and Sundays
Chandler: 'Rigmarine', South Quay, (Tel: 532657)
Repairs: A J England, (Tel: 532418). Chapman & Hewitt (Tel: 01208 813487).G B Smith & Son (Tel: 01208 862815)
Marine engineers: A&J Marine (Tel: 533114). Padstow Auto Marine (Tel: 532994). Reynolds Marine (Tel: 533033). G B Smith & Son (Tel: 01208 862815)
Electronic engineers: John Jacobsen (Tel: 01726 72501)
Transport: Buses to Bodmin to connect with rail main line. Newquay airport, flights to London
Car Parking: Large car park on quay
Yacht Clubs: Padstow Sailing Club, The Quay. Rock Sailing Club,The Quay, Rock, Wadebridge PL27 6LB (Tel: 01208 862431)
Eating out: Good pubs, cafes and restaurants in both Padstow and Rock
Things to do: Local museum and Shipwreck Museum, Padstow. Excellent beaches, swimming and walking

Visitors will lie afloat in Padstow's attractive inner harbour, which is accessible 2 hours either side of HW

PASSAGES AT A GLANCE

The following pages are intended to provide quick 'at a glance' reference to supplement the main text, arranged for the most part in the same east-west sequence

Teignmouth from the south. The Ness is a bold sandstone bluff. The bridge across the estuary can also be seen

Berry Head viewed from Torbay. Its long, flat top makes it instantly recognisable from all directions

It is not difficult to locate the Dart from the NE with its conspicuous daymark and the jagged Mewstone to seaward

Viewed from the NE Start Point is a rugged headland with a string of distinctive rocky outcrops along its spine

Passages at a glance

Start Point from the west, showing Start Rocks to seaward

Bolt Head from the SW with Great Mewstone close inshore, Little Mewstone to seaward and Prawle Point in distance

Hope Cove lies on the left of Bolt Tail, with an uncompromising stretch of coast stretching away to Bolt Head beyond

At the western end of Bigbury Bay Hillsea Point has an old lookout on its summit and Bolt Tail is visible to the east

Wembury Church is a useful mark for finding theYealm from the east. Mouthstone Point is on the extreme right

Plymouth western entrance and breakwater end.

The Draystone buoy is in the western approach to Plymouth. Rame Head, beyond, has a distinctive profile

Gribbin Head off Fowey has a prominent daymark. The Cannis Rocks are clearly marked by a buoy

Looking west the Dodman, right, is a bold headland. Nare Head and Gull Rock are in centre, Falmouth distant left

Viewed from the south-west, the Dodman is just as impressive

Passages at a glance

Falmouth from the south. St Anthony light, left. Gerrans Bay, Nare Head and Gull Rock on distant right

WHITE House

Helford River from the east. The isolated white house is useful for locating the entrance

South of Falmouth, the Manacle buoy lies well east of these dangerous rocks. Black Head is in the far distance

The Stag Rocks extend west of the Lizard . Rill Point is beyond with Asparagus Island visible against the cliffs

From a position NW of Land's End the distinctive Armed Knight rock stands clear to seaward

TIDAL STREAM CHARTS

The tidal chartlets overleaf are based on High Water Dover and only give an indication of the general direction and times of the main tidal streams. Closer inshore the streams may vary considerably in direction and strength and can often turn earlier or later than shown

Tidal stream charts

LW EXMOUTH

1 hour after H.W. Dover

02,04   01,02   04,07   03,06   06,11   11,19
02,04   03,06   04,08   09,16   06,10   07,12
03,06   05,09   09,16   06,10   07,13   15,27
10,18   07,12

LW + 1

2 hours after H.W. Dover

04,07   03,05   05,09   18,32
09,17   03,06   04,08   10,20   13,23
07,12   03,05   03,05   12,21   09,17   13,24
06,10   12,22   05,09   02,04

LW + 2

3 hours after H.W. Dover

07,12   06,11   05,09   09,16
05,09   07,13   20,36
06,10   06,10   08,15   13,23
08,15   04,08   11,20   09,16   09,16
14,25   SLACK   02,04   14,25
06,10   03,05
09,16   11,19   06,10

LW + 3

4 hours after H.W. Dover

06,11   07,12   02,03   04,07
05,09   03,06   05,09   19,34
13,24   02,03   15,27
08,15   05,09   04,08   10,18
09,16   02,04   07,13   11,19
07,13   08,15   03,06   SLACK

LW + 4

5 hours after H.W. Dover

04,08   07,12   06,11   SLACK   13,23
06,11
07,13   08,15   04,08   05,09
06,10   03,05   04,07   SLACK
04,07   07,13   06,10   09,17

LW + 5

6 hours after H.W. Dover

06,11   02,03   02,03
03,05   02,04   02,04   14,26
05,09   04,07   03,06
08,15   04,07   05,09   04,08   07,12   03,05
06,11   12,21   06,11
06,11   06,11   07,13

In offshore winds yachts can anchor off the small drying harbour of Gorran Haven in St Austell Bay

HW FALMOUTH

HW +1

HW +2

HW +3

HW +4

HW +5

5 hours before H.W. Dover

4 hours before H.W. Dover

3 hours before H.W. Dover

2 hours before H.W. Dover

1 hour before H.W. Dover

H.W. Dover

Fishcombe Cove can be a pleasant anchorage close to Brixham when conditions permit